ECOLOGY AND GLOBAL WARMING

ECOLOGY AND GLOBAL WARMING

Dr. Sarvotam Kumar

PPH

PRASHANT PUBLISHING HOUSE

NEW DELHI-110 002

First Published 2008

ISBN 978-81-906344-6-5

Published by
PRASHANT PUBLISHING HOUSE
C-34, 28 Futa Road, Gali No. 3,
West Karawal Nagar, Delhi-110 094
Ph : 011-22963970, Mob.: 9810396373

PRINTED IN INDIA

Published by Prashant Publishing House, Delhi-110 094. Typeset by Hema Typographics.

PREFACE

The issue of Global Warming is attracting World wide attention. It is feared that due to global warming, there like be wide spread floods, landslides and environmental changes, which can pore danger to a large number of people of the World Planet Earth is in real danger and the United Nations is aware of about the problem. The dangers of global warming arc more frightening to the developing world including India.

The Editor is grateful to all the American and British sources, from where researched work much of the matter has been adapted. It is expected that the book like be found to be of immense value to the students of environment, government planners sociologists teachers and the layman alike.

Editor

Contents

Contents

1

Human Role in Geomorphology

Introduction

THE HUMAN ROLE IN CREATING landforms and modifying the operation of geomorphological processes such as weathering, erosion and deposition is a theme of great importance though one that, particularly in the Western world, has not received the attention it deserves. Recently there have been some useful general surveys (see, for example, Brown, 1970; Jennings, 1966; Haigh, 1978; Nir, 1983) and, with the increasing concern for the relevance and application of geomorphological studies to human problems (see, for example, Cooke and Doornkamp, 1990; Hails, 1977), the situation is likely to change. Moreover, the development of process studies in post-Second World War geomorphology has led to a relative reduction in the importance of evolutionary and historical geomorphology within the discipline. Examination of processes, and in particular of their rate of operation, has served to highlight the part played by humans in modifying the physical landscape.

The range of the human impact on both forms and process is considerable. For example, in Table 1.1 there is a list of some anthropogenic landforms together with some of the causes of their creation. There are very few spheres of human activity which do not, even indirectly, create landforms. It is, however, useful to recognize that some features are produced by direct anthropogenic processes. These tend to be more obvious in their form and origin and are frequently created deliberately and knowingly. They include landforms produced by constructional activity (such as tipping), by excavation, by hydrological interference, and by farming (Spencer and Hale, 1961). Hillsides have been terraced in many parts of the world for many centuries, notably in the arid and semi-arid highlands of the New World (Donkin, 1979). Haigh (1978) has classified anthropogenic landforming processes as follows:

Table 1.1 Some anthropogenic landforms

Feature	*Cause*
Pits and ponds	Mining, marling
Broads	Peat extraction
Spoil heaps	Mining
Terracing, lynchets	Agriculture
Ridge and furrow	Agriculture
Cuttings	Transport
Embankments	Transport, river and coast management
Dikes	River and coast management
Mounds	Defence, memorials
Craters	War, *qanat* construction
City mounds *(tells)*	Human occupation
Canals	Transport, irrigation
Reservoirs	Water management
Subsidence depressions	Mineral and water extraction
Moats	Defence
Banks along roads	Noise abatement

1. *Direct anthropogenic processes*

   1.1 Constructional

   tipping: loose, compacted, molten
   graded: moulded, ploughed, terraced

   1.2 Excavational

   digging, cutting, mining, blasting of cohesive or non-cohesive materials
   cratered
   tramped, churned

1.3 Hydrological interference
- flooding, damming, canal construction
- dredging, channel modification
- draining
- coastal protection

2. *Indirect anthropogenic processes*

2.1 Acceleration of erosion and sedimentation
- agricultural activity and clearances of vegetation
- engineering, especially road construction and urbanization,
- incidental modifications of hydrological regime

2.2 Subsidence: collapse, settling
- mining
- hydraulic
- thermokarst

2.3 Slope failure: landslide, flow, accelerated creep
- loading, undercutting, shaking, lubrication

2.4 Earthquake generation
- loading (reservoirs)
- lubrication (fault plane).

Landforms produced by indirect anthropogenic processes are often less easy to recognize, not least because they tend to involve, not the operation of a new process or processes, but the acceleration of natural processes. They are the result of environmental changes brought about inadvertently by human technology. Nonetheless, it is probably this indirect and inadvertent modification of process and form which is the most crucial aspect of anthropogeo-morphology. By removing natural vegetation covers — through the agency of cutting, burning and grazing — humans have accelerated erosion and sedimentation. Sometimes the results will be obvious, for example when major gully systems rapidly develop; other results may have less immediate effect on landform but are, nevertheless, of great importance. By other indirect means humans may create subsidence features and problems, trigger off mass movements like landslides, and even influence the operation of phenomena like earthquakes.

Finally there are situations where, through a lack of understanding of the operation of processes and the links between different processes and phenomena, humans may deliberately and directly alter landforms and processes and thereby set in train a series of events which were not anticipated or desired. There are, for example, many records of attempts to reduce coast erosion by important and expensive engineering solutions, which, far from solving erosion problems, only exacerbated them.

Landforms produced by excavation

Of the landforms produced by direct anthropogenic processes those resulting from excavation are widespread, and may have some antiquity. For example, Neolithic peoples in the Breckland of East Anglia in England used antler picks and other means to dig a remarkable cluster of deep pits in the chalk. The purpose of this was to obtain good-quality non-frost-shattered flint to make stone tools. The cratered area, called Grimes Graves, is well displayed on aerial photographs. In many parts of Britain chalk has also been excavated to provide marl for improving acidic, light, sandy soils, and Prince (1962, 1964) has made a meticulous study of the 27,000 pits and ponds in Norfolk that have resulted mainly from this activity — an activity particularly prevalent in the eighteenth century. It is often difficult in individual cases to decide whether the depressions are the results of human intervention for solutional and periglacial depressions are often evident in the same area, but pits caused by human action do tend to have some distinctive features: irregular shape, a track leading into them, proximity to roads, etc. (see the debate between Prince and Sperling *et al.*, 1979).

Difficulties of identifying the true origin of excavational features were also encountered in explaining the Broads, a group of twenty-five freshwater lakes in the county of Norfolk. They are of sufficient area, and depth, for early workers to have precluded a human origin. Later it was proposed instead that they were natural features caused by uneven alluviation and siltation of river valleys which were flooded by the rapidly rising sealevel of the Holocene (Flandrian) transgression. It was postulated by Jennings (1952) that the Broads were initiated as a series of discontinuous natural lakes, formed beyond the limits of a thick estuarine clay wedge laid down in Romano-British times by a transgression of the sea over earlier valley peats. The waters of the Broads were thought to have been ponded back in natural peaty hollows between the flanges of the clay and the marginal valley slopes, or in tributary valleys whose mouths were blocked by the clay.

It is now clear, however, that the Broads are the result of human work (Lambert *et al.*, 1970). Some of them have rectilinear and steep boundaries, most of them are not mentioned in early topographic books, and archival records indicate that peat-cutting (turbary) was widely practised in the area. On these and other grounds it is believed that peat-diggers, before AD 1300, excavated 25.5 million m^3 of peat and so created the depressions in which the lakes have formed. The flooding may have been aided by sea-level change. Comparably extensive peat excavation was also carried on in the Netherlands, notably in the fifteenth century.

Other excavational features result from war, especially craters caused by bomb or shell impact. Regrettably, human power to create such forms is increasing. It has been calculated (Westing and Pfeiffer, 1972) that between 1965 and 1971, 26 million craters, covering an area of 1,71,000 hectares, were produced by bombing in Indo-China. This represents a total displacement of no less than 2.6 billion m^3 of earth, a figure much greater than calculated as being involved in the peaceable creation of the Netherlands.

Some excavation is undertaken on a large scale for purely aesthetic reasons, when Nature offends the eye. This can be illustrated by brief reference to the work of some English landscape gardeners. In 1676 at Cassiobury Park, Watford, Moses Cook the gardener says that he was 'forced to cut through one Hill, thirty rod, most of the hill two, feet deep, into sharp gravel', to produce the landscape he desired. At nearby Moor Park in 1720 the intended view across the Vale of St. Albans from a new house was found to be obscured by the brow of a hill, so that the hilltop was cut off by a depth of 10 m (Prince, 1959).

In many countries where land is scarce, whole hills are levelled and extensive areas stripped to provide fill for harbour reclamation. One of the most spectacular examples of this kind was the deliberate removal of steep-sided hills in the centre of Brazil's Rio de Janeiro, for housing development. In Bahrain, an island in the Arabian-Persian Gulf, a substantial part of the desert surface has been stripped to provide fill for land reclamation in the commercial and industrial area in the north of the island.

An excavational activity of a rather specialized kind is the removal of limestone pavements — areas of exposed limestone which were stripped by glaciers and then moulded into bizarre shapes by solutional activity — for ornamental rock gardens. These pavements which consist of arid, bare rock surfaces (clints) bounded by deep, humid fissures (grikes) have both an aesthetic and a biological significance (Ward, 1979). They occur in Northern England, notably in the Pennines and in the Morecombe Bay area, and provide a habitat for various rare plants (including *Actaea spicata, Ribes spicatum* and *Dynopteris villarii*). A recent survey indicates that the pavements, which only cover about 2150 hectares in Britain, have been severely damaged. On only 3 per cent of 537 pavement units recorded could no damage be detected, while the proportion considered to be 95 per cent or more intact was only 13 per cent.

The most important cause of excavation is still mineral extraction, producing open-pit mines, strip mines, quarries for structural materials, borrow pits along roads, and similar features (Doerr and Guernsely, 1956). Of these, 'without question, the environmental devastation produced by strip mining exceeds in quantity and intensity any of the other varied forms of man-made land destruction' (Strahler and Strahler, 1973:284). This form of mining is a particular environmental problem in the states of Pennsylvania, Ohio, West Virginia, Kentucky and Illinois. If, because of increasing pressure on available energy resources, the production of coal is expanded then this particular geomorphological impact may also increase. The same applies to the probable exploitation of oil shales, a potential source of oil that at present is relatively untapped. This can be exploited by open-pit mining, by traditional room and pillar mining, and by underground *in situ* pyrolysis. Vast reserves exist but the amount of excavation required will probably be about three times the amount of oil produced (on a volume basis), suggesting that the extent of both the excavation and subsequent dumping of overburden and waste will be considerable (Routson *et al.*, 1979). Some of the waste, that produced by the retorting of the shale to release the oil, contains soluble salts and potentially harmful trace elements, which limit the speed of ground reclamation (Petersen, 1981).

An attempt to provide a general picture of the importance of excavation in the creation of the landscape of Britain was given by Sherlock (1922). He estimated (see Table 1.2) that up until the time in which he wrote, human society had excavated around 31 billion m^3 of material in the pursuit of its economic activities. That figure must now be a gross underestimate, partly because Sherlock himself was not in a position to appreciate the anthropogenic role in creating features like the Norfolk Broads, and partly because, since his time, the rate of excavation has greatly accelerated. Sherlock's 1922 study covers a period when earth-moving equipment was still ill-developed. Nonetheless, on the basis of his calculations, he was able to state that 'at the present time, in a densely peopled country like England, Man is many times more powerful, as an agent of denudation, than all the atmospheric denuding forces combined.' The most notable change since Sherlock wrote has taken place in the production of aggregates for concrete which are, in descending order of importance, sand and gravel, crushed limestone, artificial or manufactured aggregates, and crushed sandstone. Demand for these materials in the UK grew from 20 million tonnes per annum in 1900, to 50 million tonnes in 1948 and 276 million tonnes in 1973, an increase in *per capita* consumption from 0.6 tonnes per year to about 5 tonnes per year (Jones, 1983).

For the world as a whole the annual movement of soil and rock resulting from mineral extraction may be as high as 3000 billion tonnes (Holdgate *et al.*, 1982: 186). By comparison it has been estimated that the amount of sediment carried into the ocean by the world's rivers each year amounts to 24 billion tonnes per year (Judson, 1968).

Landforms produced by construction and dumping

The process of constructing mounds and embankments and the creation of dry land where none previously existed is

Table 1.2 Total excavation of material in Great Britain until 1992

Activity	*Approximate volume* (m^3)
Mines	15 147 000 000
Quarries and pits	11 920 000 000
Railways	2 331 000 000
Manchester Ship Canal	41 154 000
Other canals	153 800 000
Road cuttings	480 000 000
Docks and harbours	77 000 000
Foundations of buildings and street excavation	385 000 000
Total	30 534 954 000

Source: Sherlock, 1922, p. 86

longstanding. In Britain the mound at Silbury Hill dates back to prehistoric times, and the pyramids of Central America, Egypt and the Far East are even more spectacular early feats of landform creation. Likewise hydrological management has involved, over many centuries, the construction of massive banks and walls; the ultimate result being the landscape of the present-day Netherlands. Transport developments have also required the creation of large constructional landmarks, but probably the most important features are those resulting from the dumping of waste materials, especially those derived from mining. It has been calculated that there are at least 2000 million tonnes of shale lying in pit heaps in the coalfields of Britain (Richardson, 1976). In the Middle East and other areas of long-continued human urban settlement the accumulated debris of life has gradually raised the level of the land surface, and occupation mounds (tells) are a fertile source of information to the archaeologist. Today, with the technical ability to build that humans have, even estuaries may be converted from ecologically productive environments into suburban sprawl by the processes of dredging and filling. Indeed one of the striking features of the distribution of the world's population is the tendency for large human concentrations to occur near vast expanses of water. Of those cities with more than 4 million inhabitants, more than three quarters are situated on a lake or ocean shore. Many of these cities have extended out on to land that has been reclaimed from the sea, thereby providing valuable sites for development, but sometimes causing the loss of rich fishing grounds and ecologically valuable wetlands (Hudson, 1979).

The ocean floors are also being affected because of the vast bulk of waste material that humankind is creating. Waste solid disposal by coastal cities is now sufficiently large to modify shorelines, and it covers adjacent ocean bottoms with characteristic deposits on a scale large enough to be geologically significant. This has been brought out dramatically by Gross (1972: 3174), who has undertaken a quantitative comparison of the amount of solid wastes dumped into the Atlantic by humans in the New York metropolitan region with the amount of sediment brought into the ocean by rivers:

> The discharge of waste solids exceeds the suspended sediment load of any single river along the U.S. Atlantic coast. Indeed, the discharge of wastes from the New York Metropolitan region is comparable to the estimated suspended-sediment yield (6.1 megatons per year) of all rivers along the Atlantic coast between Maine and Cape Hatteras, North Carolina.

Not only are the rates of sedimentation high, but the anthropogenic sediments tend to contain abnormally high contents of such substances as carbon and heavy metals (Goldberg *et al.*, 1978) (Table 1.3).

Many of the features created by excavation in one generation are filled in by another, since phenomena like water-filled hollows produced by mineral extraction are often both wasteful of land and suitable locations for the receipt of waste. The same applies to natural hollows such as karstic or ground-ice depressions. Watson (1976), for example, has mapped the distribution of hollows, which were largely created by marl diggers in the lowlands of South-West Lancashire and North-West Cheshire, as they were represented on mid-nineteenth-century topographic maps. When this distribution is compared with the present-day distribution for the same area, it is evident that a very substantial proportion of the holes have been infilled and obliterated by humans, with only 2114 out of 5380 remaining. Hole densities have fallen from 121 km^{-2} to 47 km^{-2}.

Table 1.3 Fluxes of heavy metals to coastal environments in the USA ($\mu g\ cm^{-2}\ y^{-1}$)

Chemical	*Santa Barbara Basin California*	*Narangansett Bay Rhode Island*	*Chesapeake Bay Cove 1411*	*Chesapeake Bay Cove 1314*
Lead				
anthropogenic	2.1	124	7.6	1.9
natural	1	2.6	5.3	3.9
Zinc				
anthropogenic	2.2	230	34	4.1
natural	9.7	14	19	15
Copper				
anthropogenic	1.4	193	4.5	0.8
natural	2.6	3.1	4	3.6

Source: Goldberg *et al.*, 1978.

Accelerated sedimentation

An inevitable consequence of the accelerated erosion produced by human activities has been accelerated sedimentation. This has been heightened by the deliberate addition of sediments to stream channels as a result of the need to dispose of mining and other wastes.

In a classic study, G.K. Gilbert (1917) demonstrated that hydraulic mining in the Sierra Nevada mountains of California led to the addition of vast quantities of sediments into the river valleys draining the range. This in itself raised their bed levels, changed their channel configurations and caused the flooding of lands that had previously been immune. Of even greater significance was the fact that the rivers transported increased quantities of debris into the estuarine bays of the San Francisco system, and caused extensive shoaling which in turn diminished the tidal prism of the bay. Gilbert calculated the volume of shoaling produced by hydraulic mining since the discovery of gold to be 846 million

m^3. Comparably serious sedimentation of bays and estuaries has also been caused by human activity on the eastern coast of America. As Gottschalk (1945: 219) wrote:

> Both historical and geological evidence indicates that the preagricultural rate of silting of eastern tidal estuaries was low. The history of sedimentation of ports in the Chesapeake Bay area is an epic of the effects of uncontrolled erosion since the beginning of the wholesale land clearing and cultivation more than three centuries ago.

He has calculated that at the head of the Chesapeake Bay, 65 million m^3 of sediment were deposited between 1846 and 1938. The average depth of water over an area of 83 km^2 was reduced by 0.76 m. New land comprising 318 hectares was added to the state of Maryland and, as Gottschalk remarked, 'the Susquehanna River is repeating the history of the Tigris and Euphrates'. Much of the material entrained by erosive processes on upper slopes as a result of agriculture in Maryland, however, was not evacuated as far as the coast. Costa (1975) has suggested, on the basis of the study of sedimentation, that only about one-third of the eroded material left the river valleys. The remainder accumulated on floodplains as alluvium and colluvium at rates of up to 1.6 cm per year. Similarly, Happ (1944), working in Wisconsin, carried out an intensive augering survey of floodplain soil and established that, since the development of agriculture, floodplain aggradation had proceeded at a rate of approximately 0.85 cm per year. He noted that channel and floodplain aggradation had caused the flooding of low alluvial terraces to be more frequent, more extensive and deeper. The rate of sedimentation has since declined (Trimble, 1976).

Such valley sedimentation is by no means restricted to the newly settled terrains of North America. There is increasing evidence to suggest that silty valley fills in Germany, France and Britain, many of them dating back to the Bronze Age and the Iron Age, are the result of accelerated slope erosion produced by the activities of early farmers (Bell, 1982). Indeed, in recent years, various studies have been undertaken with a view to assessing the importance of changes in sedimentation rate caused by humans at different times in the Holocene in Britain. Among the formative events that have been identified are initial land clearance by Mesolithic and Neolithic peoples; agricultural intensification and sedentarization in the late Bronze Age; the widespread adoption of the iron plough in the early Iron Age; settlement by the Vikings and the introduction of sheep farming (Table 1.4).

Table 1.4 Accelerated sedimentation in Britain in prehistorical and historical times

Location	*Source*	*Evidence and date*
Howgill Fells	Harvey *et al.* (1981)	Debris cone production following tenth century AD introduction of sheep farming
Upper Thames basin	Robinson and Lambrick (1984)	River alluviation in Late Bronze Age and early Iron Age
Lake District	Pennington (1981)	Accelerated lake sedimentation at 5000 BP as a result of Neolithic agriculture
Mid-Wales	Macklin and Lewin (1986)	Floodplain sedimentation (Capel Bangor unit) on Rheidol as a result of early Iron Age sedentary agriculture
Brecon Beacons	Jones *et al.* (1985)	Lake sedimentation increase after 5000 BP at Llangorse due to forest clearance
Weald	Burrin (1985)	Valley alluviation from Neolithic onwards until early Iron Age
Bowland Fells	Harvey and Renwick (1987)	Valley terraces at 5000–2000 BP (Bronze or Iron Age settlement) and after 1000 BP (Viking settlement)
Southern England	Bell (1982)	Fills in dry valleys Bronze Age and Iron Age
Callaly Moor (Northumberland)	Macklin *et al.* (1991)	Valley fill sediments of late Neolithic to Bronze Age

A core from Llangorse Lake in the Brecon Beacons of Wales (Jones *et al.*, 1985) provides excellent long-term data on changing sedimentation rates:

Period (years BP)	*Sedimentation rate* *(cm 100 y^{-1})*
9000–7500	3.5
7500–5000	1.0
5000–2800	13.2
2800–AD 1840	14.1
*c.*AD 1840–present	59.0

The thirteenfold increase in rates after 5000 BP seems to have occurred rapidly and is attributed to initial forest clearance. The second dramatic increase of more than fourfold took place in the last 150 years and is a result of agricultural intensification.

The work of Binford *et al.* (1987) on the lakes of the Peten region of Northern Guatemala (Central America), an area of tropical lowland dry forest, is also instructive with respect to early agricultural colonization. Combining studies of archaeology and lake sediment stratigraphy, they were able to reconstruct the diverse environmental consequences of the growth of Mayan civilization. This civilization showed a dramatic growth after 3000 years BP, but collapsed in the ninth century AD. The hypotheses put forward to explain this collapse include warfare, disease, earthquakes and soil degradation. The population has remained relatively low ever since, and after the first European contact (AD 1525) the region was virtually depopulated. The period of Mayan success saw a marked reduction in vegetation cover, an increase in lake sedimentation rates and in catchment soil erosion, an increased supply of inorganic silts and clays to the lakes, a pulse of phosphorus derived from human wastes, and a decrease in lacustrine productivity caused by high levels of turbidity.

In the last two centuries rates of sedimentation in lake basins have changed in different ways in different basins according to the differing nature of economic activities in catchments.

Some data from various sources are listed for comparison in Table 1.5. In the case of the Loe Pool in Cornwall

South-West England rates of sedimentation were high while mining industry was active, but fell dramatically when mining was curtailed. In the case of Seeswood Pool in Warwickshire, a dominantly agricultural catchment area in Central England, the highest rates have occurred since 1978 in response to various land management changes, such as larger fields, continuous cropping and increased dairy herd size.

Table 1.5 Data on rates of erosion and sedimentation in lakes in the last two centuries

Dates	*Activity*	*Rates of erosion in catchment (R. Cober) as determined from lake sedimentation rates Activity (t km^{-2} y^{-1})*
(a) Loe Pool (Cornwall) (from O'Sullivan et al., 1982)		
1860–1920	Mining and agriculture	174
1930–36	Intensive mining and agriculture	421
1937–38	Intensive mining and agriculture	361
1938–81	Agriculture	12
(b) Seeswood Pool (Warwickshire) (from Foster et al., 1986)		
1765–1853		7.0
1854–80		12.2
1881–1902		8.1
1903–19		9.6
1920–25		21.6
1926–33		16.1
1934–47		12.7
1948–64		12.0
1965–72		13.9
1973–77		18.3
1978–82		36.2

In other catchments, preafforestation ploughing may have caused sufficient disturbance to cause accelerated sedimentation. For example, Battarbee *et al.* (1985b) looked at sediment cores in the Galloway area of South-West Scotland and found that in Loch Grannoch the introduction of ploughing in the catchment caused an increase in sedimentation from 0.2 cm y^{-1} to 2.2 cm y^{-1}.

Sedimentation has severe economic implications because of the role it plays in reducing the effective lifetime of reservoirs. In the tropics capacity depletion through sedimentation is commonly around 2 per cent per annum (Myers, 1988: 14), so that the expected useful life of the Paute hydroelectric project in Ecuador (cost US$ 600 million) is 32 years, that of the Mangla Dam in Pakistan (cost US$ 600 million) is 57 years, and that of the Tarbela Dam, also in Pakistan, is just 40 years.

Ground subsidence

Not all ground subsidence is caused by humans. For example, limestone solutional processes can, in the absence of humankind, create a situation where a cavern collapses to produce a surface depression, and permafrost will sometimes melt to produce a thermokarst depression without human intervention. Nonetheless, ground subsidence can be caused or accelerated by humans in a variety of ways: by the transfer of subterranean fluids (such as oil, gas and water); by the removal of solids through underground mining or by dissolving solids and removing them in solutions (for example, sulphur and salt); by the disruption of permafrost, and by the compaction or reduction of sediments because of drainage and irrigation (Johnson, 1991).

Some of the most dangerous and dramatic collapses have occurred in limestone areas because of the dewatering of limestone caused by mining activities. In the Far West Rand of South Africa, gold-mining has required the abstraction of water to such a degree that the local water-table has been lowered by more than 300 m. The fall of the water-table caused miscellaneous clays and other materials filling the roofs of large caves to dry out and shrink so that they collapsed into the underlying void. One collapse created a depression 30 m deep and 55 m across, killing twenty-nine people. In Alabama, USA, water-level decline consequent upon pumping has had equally serious consequences in a limestone terrain; and Newton (1976) has estimated that since 1900 about 4000 induced sinkholes or related features have been formed, while fewer than fifty natural sinkholes have been reported over the same interval. Sinkholes that may result from such human activity are also found in Georgia, Florida, Tennessee, Pennsylvania and Missouri.

In some limestone areas, however, a reverse process can operate. The application of water to overburden above the limestone may render it more plastic so that the likelihood of collapse is increased. This has occurred beneath reservoirs, such as the May Reservoir in Central Turkey, and as a result of the application of waste water and sewerage to the land surface.

The process can be accelerated by the direct solution of susceptible rocks. For example, collapses have occurred in gypsum bedrock because of solution brought about by the construction of a reservoir. In 1893 the MacMillan Dam was built on the Pecos River in New Mexico, but within twelve years the whole river flowed through caves which had developed since construction. Both the San Fernando and Rattlesnake Dams in California suffer severe leakage for similar reasons.

Subsidence produced by oil abstraction is an increasing problem in some parts of the world. The classic area is Los Angeles, where 9.3 m of subsidence occurred as a result of exploitation of the Wilmington oilfield between 1928 and 1971. The Inglewood oilfield displayed 2.9 m of subsidence between 1917 and 1963. Some coastal flooding problems occurred at Long Beach because of this process. Similar subsidence has been recorded from the Lake Maracaibo field in Venezuela (Prokopovich, 1972) and from some Russian fields (Nikonov, 1977).

A more widespread problem is posed by ground-water abstraction for industrial, domestic and agricultural purposes. Table 1.6 presents some data for such subsidences from various parts of the world. The ratios of subsidence to water-level decline are strongly dependent on the nature of the sediment composing the aquifer, so that ratios range from 1:7 for Mexico City, to 1:80 for the Pecos in Texas, and to less than 1:400 for London, England (Rosepiler and Reilinger, 1977).

The extent of subsidence that has taken place in the United States as a result of groundwater abstraction has recently been

Table 1.6 Ground subsidence

Location	*Amount (m)*		*Rate (mmly)*
(a) Ground subsidence produced by oil and gas abstraction			
Azerbaydzhan, USSR	2.5	(1912–62)	50
Atravopol, USSR	1.5	(1956–62)	125
Wilmington, USA	9.3	(1928–71)	216
Inglewood, USA	2.9	(1917–63)	63
Maracaibo, Venezuela	5.03	(1929–90)	84
(b) Ground subsidence produced by groundwater abstraction			
London, England	0.06–0.08	(1865–1931)	0.91–1.21
Savannah, Georgia (USA)	0.1	(1918–55)	2.7
Mexico City	7.5	–	250–300
Houston, Galveston, Texas	1.52	(1943–64)	60–76
Central Valley, California	8.53	–	–
Tokyo, Japan	4	(1892–1972)	500
Osaka, Japan	>2.8	(1935–72)	76
Niigata, Japan	>1.5	–	–
Pecos, Texas	0.2	(1935–66)	6.5
South-central Arizona	2.9	(1934–77)	96
Bangkok, Thailand	0.5	–	100
Shanghai, China	2.62	(1921–65)	60

Source: data in Cooke and Doornkamp, 1974; Prokopovich, 1972; Rosepiler and Reilinger, 1977; Holzer, 1979; Nutalaya and Rau, 1981; Johnson, 1991.

reassessed on the basis of geodetic evidence (Chi and Reilinger, 1984). To the major areas previously identified:

South-central Arizona
Savannah (Georgia)
Pecos and Houston—Galveston (Texas)
Denver (Colorado)
San Joaquin Valley, Santa Clara Valley, Saugus Basin, Los Angeles Basin, Bunker Hill–San Timoteo (California)
Milford (Utah)
Raft River Valley (Idaho)
Baton Rouge and New Orleans (Louisiana)
Las Vegas (Nevada)

they have added:

> Ventura, Ontario and San Pedr–Santa Monica (California)
> Monroe and Alexandria (Louisiana)
> Jackson (Mississippi).

In Japan subsidence has now emerged as a major problem (Nakano and Matsuda, 1976). In 1960 only 35.2 km^2 of the Tokyo lowland was below sealevel, but continuing subsidence meant that by 1974 this had increased to 67.6 km^2, exposing a total of 1.5 million people to major flood hazard.

The subsidence caused by mining (which led to court cases in England as early as the fifteenth century as a consequence of associated damage to property) is perhaps the most familiar, though its importance varies according to such factors as the thickness of seam removed, its depth, the width of working, the degree of filling with solid waste after extraction, the geological structure, and the method of working adopted (Wallwork, 1974). In general terms, however, the vertical displacement by subsidence is less than the thickness of the seam being worked, and decreases with an increase in the depth of mining. This is because the overlying strata collapse, fragment and fracture, so that the mass of rock fills a greater space than it did when naturally compacted. Consequently the surface expression of deep-seated subsidence may be equal to little more than one-third of the thickness of the material removed. Subsidence associated with coal-mining may disrupt surface drainage and the resultant depressions then become permanently flooded. In England, the valleys of the Aire and Calder in the Castleford-Knottingley area, the Anker Valley in East Warwickshire, and the Wigan – Leigh area of Lancashire, provide examples of subsidence lakes formed on stream floodplains.

Coal-mining regions are not the only areas where subsidence problems are serious. In Cheshire, North-West England, salt is extracted from two major seams, each about 30 m in thickness. Moreover, these seams occur at no great depth—the uppermost being at about 70 m below the surface. A further factor to be considered is that the rock salt is highly soluble in water, so the flooding of mines may cause additional collapse. These three conditions – thick seams, shallow depth and high solubility—have produced optimum conditions for subsidence and many subsidence lakes called 'flashes' have developed (Wallwork, 1956). The main railway line between Crewe and Manchester at Elton subsided no less than 4.9 m between 1892 and 1956 (Wallwork, 1960). Likewise, salt extraction at Windsor, Canada, is known to have produced a water-filled depression 150 m in diameter and about 8 m deep (Martinez, 1971).

Some subsidence is created by a process called hydrocompaction, which is explained thus. Moisture-deficient, unconsolidated, low-density sediments tend to have sufficient dry strength to support considerable effective stresses without compacting. However, when such sediments, which may include alluvial fans or loess, are thoroughly wetted for the first time (for example, by percolating irrigation water) the inter-granular strength of the deposits is diminished, rapid compaction takes place, and ground surface subsidence follows. Unequal subsidence can create problems for irrigation schemes.

Land drainage can promote subsidence of a different type, notably in areas of organic soils. The lowering of the watertable makes peat susceptible to oxidation and deflation so that its volume decreases. One of the longest records of this process, and one of the clearest demonstrations of its efficacy, has been provided by the measurements at Holme Fen Post in the English Fenlands. Approximately 3.8 m of subsidence occurred between 1848 and 1957 (Fillenham, 1963), with the fastest rate occurring soon after drainage had been initiated. The present rate averages about 1.4 cm y^{-1} (Richardson and Smith, 1977). At its maximum natural extent the peat of the English Fenland covered around 1750 km^2. Now only about one-quarter (430 km^2) remains.

A further type of subsidence, sometimes associated with earthquake activity, results from the effects on the earth's crust of large masses of water impounded behind reservoirs. Seismic effects can be generated in areas with susceptible fault systems and this may account for earthquakes recorded at Koyna (India) and elsewhere. The process whereby a mass of water causes coastal depression is called hydro-isostasy, and the degree of subsidence that has been monitored for various large reservoirs is listed in Table 1.7.

Table 1.7 Hydro-isostatic subsidence caused by the creation of reservoirs

Reservoir	*Maximum downwarping (cm)*	*Period*
Lake Mead, USA	20.1	1950–63
Kariba, Central Africa	12.7	1959–68
Koyna, India	8–14	1962–68
Bratsk, Siberia	5.6	1961–66
Krasnoyarsk, Siberia	3	1967–71
Plyava, Baltic	0.5	1965–70

Source: from data in Nikonov, 1977.

In permafrost areas ground subsidence is associated with thermokarst development, thermokarst being irregular, hummocky terrain produced by the melting of ground ice, permafrost. The development of thermokarst is due primarily to the disruption of the thermal equilibrium of the permafrost and an increase in the depth of the active layer. Following French (1976: 106), consider an undisturbed tundra soil with an active layer of 45 cm. Assume also that the soil beneath 45 cm is supersaturated permafrost and yields on a volume basis upon thawing 50 per cent excess water and 50 per cent saturated soil. If the top 15 cm were removed, the equilibrium thickness of the active layer, under the bare ground conditions, might increase to 60 cm. As only 30 cm of the original active layer remains, 60 cm of the permafrost must thaw before the active layer can thicken to 60 cm, since 30 cm of supernatant water will be released. Thus, the surface subsides 30 cm because of thermal melting associated with the degrading permafrost, to give an overall depression of 45 cm.

Thus the key process involved in thermokarst subsidence is the state of the active layer and its thermal relationships. When, for example, surface vegetation is cleared for agricultural or constructional purposes the depth of thaw will tend to increase. The movement of tracked vehicles has been particularly harmful to surface vegetation and deep channels may soon result from permafrost degradation. Similar effects may be produced by the siting of heated buildings on permafrost, and by the laying of oil, sewer and water pipes in or on the active layer (Ferrians *et al.*, 1969; Lawson, 1986).

Thus subsidence is a diverse but significant aspect of the part humans play as geomorphological agents. The damage caused on a worldwide basis can be measured in billions of dollars each year (Coates, 1983), and among the effects are broken dams, cracked buildings, offset roads and railways, fractured well casings, deformed canals and ditches, bridges that need relevelling, saline encroachment and increased flood damage.

Arroyo trenching, gullies and peat haggs

In the South-western United States many broad valleys and plains became deeply incised with valley-bottom gullies (arroyos) over a short period between 1865 and 1915, with the 1880s being especially important (Cooke and Reeves, 1976). This cutting had a rapid and detrimental effect on the flat, fertile and easily irrigated valley floors, the most desirable sites for settlement and economic activity in a harsh environment.

Many students of this phenomenon have believed that thoughtless human actions caused the entrenchment, and the apparent coincidence of white settlement and arroyo development tended to give credence to this viewpoint. The range of actions that could have been culpable is large: timber-felling, overgrazing, cutting grass for hay in valley bottoms, compaction along well-travelled routes, channelling of run-off from trails and railways, disruption of valley-bottom sods by animals' feet, and the invasion of grasslands by miscellaneous types of scrub.

On the other hand, study of the long-term history of the valley fills shows that there have been repeated phases of aggradation and incision and that some of these took place before the influence of humans could have been a significant factor. This has prompted debate as to whether the arrival of white communities was in fact responsible for this particularly severe phase of environmental degradation. Huntington (1914), for example, argued that valley filling would be a consequence of a climatic shift to more arid conditions. These, he believed, would cause a reduction in vegetation which in turn would promote rapid removal of soil from devegetated mountain slopes during storms, and overload streams with sediment.

With a return to humid conditions vegetation would be re-established, sediment yields would be reduced and entrenchment of valley fills would take place. Bryan (1928) put forward a contradictory climatic explanation. He argued that a slight move towards drier conditions, by depleting vegetation cover and reducing soil infiltration capacity, would produce significant increases in storm run-off which would erode valleys. Another climatic interpretation was advanced by Leopold (1951), involving a change in rainfall intensity rather than quantity. He indicated that a reduced frequency of low-intensity rains would weaken the vegetation cover, while an increased frequency of heavy rains at the same time would increase the incidence of erosion. Support for this contention comes from the work of Balling and Wells (1990) in New Mexico. They attributed early twentieth-century arroyo trenching to a run of years with intense and erosive rainfall characteristics that succeeded a phase of drought conditions in which the productive ability of the vegetation had declined.

It is also possible, as Schumm *et al.* (1984) have pointed out, that arroyo incision could result from neither climatic change nor human influence. It could be the result of some natural geomorphological threshold being crossed.

It is therefore clear that the possible mechanisms that can lead to alternations of cut-and-fill of valley sediments are extremely complex, and that any attribution of all arroyos in all areas to human activities may be a serious oversimplification of the problem. In addition, it is possible that natural environmental changes, such as changes in rainfall characteristics, have operated at the same time and in the same direction as human actions.

In the Mediterranean lands there have also been controversies surrounding the age and causes of alternating phases of aggradation and erosion in valley bottoms. Vita-Finzi (1969) has suggested that at some stage during historical times many of the streams in the Mediterranean area, which had hitherto been engaged primarily in downcutting, began to build up their beds. Renewed downcutting, still seemingly in operation today, has since incised the channels into the alluvial fill. He proposed that the reversal of the downcutting trend in the Middle Ages was both ubiquitous and confined in time, and that some universal and time-specific agency was required to explain it. He believed that devegetation by humans was not a medieval innovation and that some other mechanism was required. A solution he gave to account for the phenomenon was precipitation change during the climatic fluctuation known as the Little Ice Age (AD 1550–1850). This was not an interpretation which found favour with Butzer (1974). He reported that his investigations showed plenty of post-Classical and pre-1500 alluviation (which could not therefore be ascribed to the Little Ice Age), and he doubted whether Vita-Finzi's dating was precise enough to warrant a 1550–1850 date. Instead, he suggested that humans were responsible for multiple phases of accelerated erosion from slopes, and accelerated sedimentation in valley bottoms, from as early as the middle of the first millennium BC.

Butzer's interpretation has found favour with van Andel *et al.* (1990) who have detected an intermittent and complex record of cut and fill episodes during the late Holocene in various parts of Greece. They believe that this evidence is compatible with a model of the control of timing and intensity of landscape destabilization by local economic and political conditions. This is a view shared in the context of the Algarve in Portugal by Chester and James (1991).

Another example of drainage incision that demonstrates the problem of disentangling the human from the natural causes of erosion is provided by the eroding peat bogs of highland Britain. Over many areas, including the Pennines of Northern England and the Brecon Beacons of Wales, blanket peats are being severely eroded to produce pool and hummock topography, areas of bare peat and incised gullies (haggs). Many rivers draining such areas are discoloured by the presence of eroded peat particles, and sediment yields of organic material are appreciable (Labadz *et al.*, 1991).

Some of the observed peat erosion may be an essentially natural process, for the high water content and low cohesion of undrained peat masses make them inherently unstable. Moreover, the instability must normally become more pronounced as peat continues to accumulate, leading to bog slides and bursts round margins of expanded peat blankets. Conway (1954) suggested that an inevitable endpoint of peat build-up on high altitude, flat or convex surfaces is that a considerable depth of unconsolidated and poorly humidified peat overlies denser and well-humidified peat, so adding to the instability. Once a bog burst or slide occurs, this leads to the formation of drainage gullies which extend back into the peat mass, slumping-off marginal peat downslope, and leading to the drawing off of water from the pools of the hummock and hollow topography of the watershed.

Tallis (1985a) believes that there have been two main phases of erosion in the Pennines. The first, initiated 1000–1200 years ago, may have been caused by natural instability of the type outlined above. However, there has been a second stage of erosion, initiated 200–300 years ago, in which miscellaneous human activities appear to have been important. Radley (1962) suggested that among the pressures that had caused erosion were heavy sheep grazing, regular burning, peat cutting, the digging of boundary ditches, the incision of pack horse tracks, and military manoeuvres during the First World War. Other causes may include footpath erosion and severe air pollution (Tallis, 1965). In South Wales, there is some evidence that the blanket peats have degenerated as a result of contamination by particulate pollution (soot, etc.) during the Industrial Revolution (Chambers *et al.*, 1979). On the other hand, in Scotland lake core studies indicate that severe peat erosion was initiated between AD 1500 and 1700, prior to air pollution associated with industrial growth, and Stevenson *et al.* (1990) suggest that this erosion initiation may have been caused either by the adverse climatic conditions of the Little Ice Age or by an increasing intensity of burning as land use pressures increased.

Accelerated weathering and the tufa decline

Although fewer data are available and the effects are generally less immediately obvious, there is some evidence that human activities have produced changes in the nature and rate of weathering (Winkler, 1970). The prime cause of this is probably air pollution. It is clear that, as a result of increased emissions of sulphur dioxide through the burning of fossil fuels, there are higher levels of sulphuric acid in rain over many industrial areas. This in itself may react with stones and cause their decay. Chemical reactions involving sulphur dioxide can also generate salts such as calcium sulphate and magnesium sulphate, which may be effective in causing the physical breakdown of rock through the mechanism of salt weathering.

Similarly, atmospheric carbon-dioxide levels have been rising steadily because of the burning of fossil fuels, and deforestation. Carbon dioxide may combine with water, especially at lower temperatures, to produce weak carbonic acid which can dissolve limestone, marbles and dolomites. Weathering can also be accelerated by changes in groundwater levels resulting from irrigation. This can be illustrated by considering the Indus Plain in Pakistan (Goudie, 1977), where irrigation has caused the watertable to be raised by about 6 m since 1922. This has produced increased evaporation and salinization. The salts that are precipitated by evaporation above the capillary fringe include sodium sulphate, a very effective cause of stone decay. Indeed buildings, such as the great archaeological site of Mohenjo-Daro, are decaying at a catastrophic rate.

In other cases accelerated weathering has been achieved by moving stone from one environment to another. Cleopatra's Needle, an Egyptian obelisk in New York City, is an example of rapid weathering of stone in an inhospitable environment. Originally erected on the Nile opposite Cairo about 1500 BC, it was toppled in about 500 BC by Persian invaders, and lay partially buried in Nile sediments until, in 1880, it was moved to New York. It immediately began to suffer from scaling and the inscriptions were largely obliterated within ten years because of the penetration of moisture which enabled the frost-wedging and hydration of salts to occur.

During the 1970s and 1980s an increasing body of isotopic dates became available for deposits of tufa (secondary freshwater deposits of limestone, also known as travertine). Some of these dates suggest that over large parts of Europe, from Britain to the Mediterranean basin and from Spain to Poland, rates of tufa formation were high in the early and mid-Holocene, but declined markedly thereafter (Weisrock, 1986: 165-67). Vaudour (1986) maintains that since around 3000 BP 'the impact of man on the environment has liberated their disappearance,' but he gives no clear indication of either the basis of this point of view or of the precise mechanism(s) that might be involved. If the late

Holocene reduction in tufa deposition is a reality then it is necessary to consider a whole range of possible mechanisms, both natural and anthropogenic (Nicod, 1986: 71–80; Table 1.8). As yet the case for an anthropogenic role is not proven.

Table 1.8 Some possible mechanisms to account for the alleged Holocene tufa decline

Climatic/natural	*Anthropogenic*
Discharge reduction following rainfall decline leading to less turbulence	Discharge reduction due to overpumping. diversions, etc.
Degassing leads to less deposition	Increased flood scour and run-off variability of channels due to deforestation, urbanization, ditching, etc.
Increased rainfall causing more flood scour	Channel shifting due to deforestation of flood plains leads to tufa erosion
Decreasing temperature leads to less evaporation and more CO_2 solubility	Reduces CO_2 flux in system after deforestation
Progressive Holocene peat development and soil podzol development through time leads to more acidic surface waters	Introduction of comestic stock causes breakdown of fragile tufa structures Deforestation = less fallen trees to act as foci for tufa barrages Increase stream turbidity following deforestation reduces algal productivity

Accelerated mass movements

There are many examples of mass movements being triggered by human actions (Selby, 1979). For instance, landslides can be created either by undercutting or by overloading. When a road is constructed, material derived from undercutting the upper hillside may be cast on to the lower hillslope as a relatively loose fill to widen the road bed. Storm water is then often diverted from the road on to the loose fill.

Because of the hazards presented by both natural and accelerated mass movements, humans have developed a whole series of techniques to attempt to control them. Such methods, many of which are widely used by engineers, are listed in Table 1.9. These techniques are increasingly necessary, as human capacity to change a hillside and to make it more prone to slope failure has been transformed by engineering development. Excavations are going deeper, buildings and other structures are larger, and many sites which are at best marginally suitable for engineering projects are now being used because of increasing pressure on land. This applies especially to some of the expanding urban areas in the humid parts of low latitudes—Hong Kong, Kuala Lumpur, Rio de Janeiro and many others. It is very seldom that human agency deliberately accelerates mass movements; most are accidentally caused, the exception possibly being the deliberate triggering of a threatening snow avalanche (Perla, 1978).

The forces producing slope instability and landsliding can usefully be divided into disturbing forces and resisting properties (Cooke and Doornkamp, 1990: 113–14). The factors leading to an increase in shear stress (disturbing forces) can be listed as follows (modified after Cooke and Doornkamp, 1990: 113):

1. *Removal of lateral or underlying support*
   Undercutting by water (for example, river, waves), or glacier ice
   Weathering of weaker strata at the toe of the slope
   Washing out of granular material by seepage erosion
   human cuts and excavations, drainage of lakes or reservoirs.
2. *Increased disturbing forces*
   Natural accumulations of water, snow, talus
   pressure caused by human activity (for example, stockpiles of ore, tip-heaps, rubbish dumps, or buildings).
3. *Transitory earth stresses*
   Earthquakes
   Continual passing of heavy traffic
4. *Increased internal pressure*
   Build up of pore-water pressures (for example, in joints and cracks, especially in the tension crack, zone at the rear of the slide).

Table 1.9 Examples of methods of controlling mass movements

Type of movement	*Method of control*
Falls	Flattening the slope Benching the slope Drainage Reinforcement of rock walls by grouting with cement, anchor bolts Covering of wall with steel mesh
Slides and flows	Grading or benching to flatten the slope Drainage of surface water with ditches Sealing surface cracks to prevent infiltration Sub-surface drainage Rock or earth buttresses at foot Retaining walls at foot Pilings through the potential slide mass

Source: after R. F. Baker and H. E. Marshall, in Dunne and Leopold, 1978.

Some of the factors are natural, while others (italicized) are effected by humans. Factors leading to a decrease in the shearing resistance of materials making up a slope can also be summarized (also modified after Cooke and Doornkamp, 1990: 113):

1. *Materials*

   Beds which decrease in shear strength if water content increases (clays, shale, mica, schist, talc, serpentine) (for example, *when local watertable is artificially increased in height by reservoir construction*), or as a result of stress release (vertical and/or horizontal) following slope formation. Low internal cohesion (for example, consolidated clays, sands, porous organic matter).

   In bedrock: faults, bedding planes, joints, foliation in schists, cleavage, brecciated zones, and pre-existing shears.

2. *Weathering changes*

   Weathering reduces effective cohesion, and to a lesser extent the angle of shearing resistance

Absorption of water leading to changes in the fabric of clays (for example, loss of bonds between particles or the formation of fissures).

3. *Pore-water pressure increase*

High groundwater table as a result of increased precipitation or *as a result of human interference* (for example, *dam construction)* (see 1 above).

Once again the italics show that there are a variety of ways in which humans can play a role.

Some mass movements are created by humans piling up waste soil and rock into unstable accumulations that fail spontaneously. At Aberfan, in South Wales, a major disaster occurred when a 180 m high coal-waste tip began to move as an earth flow. The tip had been constructed not only as a steep slope but also upon a spring line. This made an unstable configuration which eventually destroyed a school and claimed over 150 lives. Human-induced mass movements are also a problem in Los Angeles, California. Some result from the denudation of slope vegetation cover by fires, many of which are set by humans, while others are triggered by infiltration of water from cesspools and from irrigation water applied to lawns and gardens. In Hong Kong, where a large proportion of the population is forced to occupy steep slopes developed on deeply weathered granites and other rocks, mass movements are a severe problem, and So (1971) has shown that many of the landslips and washouts (70 per cent of those in the great storm of June 1966, for example) were associated with road sections and slopes artificially modified through construction and cultivation.

One of the most serious mass movements partly caused by human activity was that which caused the Vaiont Dam disaster in Italy in 1963, in which 2600 people were killed (Kiersch, 1965). Heavy antecedent rainfall conditions and the presence of young, highly folded sedimentary rocks provided the necessary conditions for a slip to take place, but it was the construction of the Vaiont Dam itself which changed the local groundwater conditions sufficiently to affect the stability of a rock mass on the margins of the reservoir: 240 million m^3 of ground slipped with enormous speed into the reservoir, producing a sharp rise in water level which overtoppped the dam and caused flooding and loss of life downstream. Comparable slope instability resulted when the Franklin D. Roosevelt lake was impounded by the Columbia River in the USA (Coates, 1977), but the effects were, happily, less serious.

It is evident from what has been said about the predisposing causes of the slope failure *triggered* by the Vaiont Dam that human agency was only able to have such an impact because the natural conditions were broadly favourable. Exactly the same lesson can be learnt from the accelerated landsliding in Southern Italy. Nossin (1972) has demonstrated how, in Calabria, road construction has triggered off (and been hindered by) landsliding, but he has also stressed that the area is fundamentally susceptible to such mass movement activity because of geological conditions. It is an area where recent rapid uplift has caused downcutting by rivers and the undercutting of slopes by erosion. It is also an area of incoherent metamorphic rocks, with frequent faulting. Further, water is often trapped by Tertiary clay layers, providing further stimulus to movement.

Although the examples of accelerated mass movements that have been given here are essentially associated with the effects of modern construction, projects, more long-established activities, including deforestation and agriculture, are also highly important. For example, Innes (1983) has demonstrated, on the basis of licheno-metric dating of debris-flow deposits in the Scottish Highlands, that most of the flows have developed in the last 250 years, and he suggests that intensive burning and grazing may be responsible.

However, as with so many environmental changes of that nature in the past, there are considerable difficulties in being certain about causation. This has been well expressed by Ballantyne (1991: 84):

> Although there is growing evidence for Late Holocene erosion in upland Britain, the causes of this remain elusive. A few studies have presented evidence linking erosion to vegetation degradation and destruction due to human influence, but the validity of climatic deterioration as a cause of erosion remains unsubstantiated. This uncertainty stems from a tendency to link erosion with

particular causes only through assumed coincidence in timing, a procedure fraught with difficulty because of imprecision in the dating of both putative causes and erosional effects. Indeed, in many reported instances it is impossible to refute the possibility that the timing of erosional events or episodes may be linked to high magnitude storms of random occurrence, and bears little relation to either of the causal hypotheses outlined above...

Deliberate modification of channels

Both for purposes of navigation and flood control humans have deliberately straightened many river channels. Indeed, the elimination of meanders contributes to flood control in two ways. First, it eliminates some overbank floods on the outside of curves, against which the swiftest current is thrown and where the water surface rises highest. Second, and more importantly, the resultant shortened course increases both the gradient and the velocity, and the floodwaters erode and deepen the channel, thereby increasing its flood capacity.

It was for this reason that a programme of channel cut-offs was initiated along the Mississippi in the early 1930s. By 1940 it had lowered flood stages by as much as 4 m at Arkansas City, Arkansas. By 1950 the length of the river between Memphis, Tennessee and Baton Rouge, Louisiana (600 km down the valley) had been reduced by no less than 270 km as a result of 16 cut-outs.

The largest current scheme for channel modification is the Jonglei Canal in the Sudan. This scheme has involved the construction of a channel 360 km long that will bypass the Sudd swamps where more than half of the White Nile's flow is lost by evapotranspiration every year. This will provide an extra 4.75 billion m^3 of water each year for Sudan and Egypt, but its impact on the life of the local pastoral people and on the ecology of the flood plains is the cause of considerable misgivings (Charnock, 1983).

Some landscapes have become dominated by artificial channels, normally once again because of the need for flood alleviation and drainage. This is especially evident in an area like the English Fenlands where straight constructed channels contrast with the sinuous courses of original rivers such as the Great Ouse.

Non-deliberate river-channel changes

There are thus many examples of the intentional modification of river-channel geometry by humans, by the construction of embankments, by channelization, and by other such processes. However, major changes in the configuration of channels can be achieved accidentally (Table 1.10), either because of human-induced changes in stream discharge or in their sediment load: both parameters affect channel capacity (Park, 1977).

Deliberate channel-straightening causes various types of sequential channel adjustment both within and downstream from straightened reaches, and the types of adjustment vary according to such influence as stream gradient and sediment characteristics. Brookes (1987) has recognized five types of change *within* the straightened reaches (types Wl to W5) and two types of change downstream (types Dl and D2).

Table 1.10 Accidental channel changes

Phenomenon	*Cause*
Channel incision	'Clear-water erosion' below dams caused by sediment removal
Channel aggradation	Reduction in peak flows below dams Addition of sediment to streams by mining, agriculture, etc.
Channel enlargement	Increase in discharge level produced by urbanization
Channel diminution	Discharge decrease following water abstraction or flood control
Channel diminution	Trapping and stabilizing of sediment by artificially introduced plants

Type W1 is degradation of the channel bed, which results from the fact that straightening increases the slope by providing a shorter channel path. This in turn increases its sediment transport capability.

Type W2 is the development of an armoured layer on the channel bed by the more efficient removal of fine materials as a result of the increased sediment transport capability referred to above.

Type W3 is the development of a sinuous thalweg in streams which are not only straightened but which are also widened beyond the width of the natural channel.

Type W4 is the recovery of sinuously as a result of bank erosion in channels with high slope gradients.

Type W5 is the development of a sinuous course by deposition in streams with a high sediment load and a relatively low valley gradient.

Types D1 and D2 result from deposition downstream as the stream tries to even out its gradient, the deposition occurring as a general raising of the bed level, or as a series of accentuated point bar deposits.

It is now widely recognized that the urbanization of a river basin results in an increase in the peak flood flows in a river. It is also recognized that the morphology of stream channels is related to their discharge characteristics, and especially to the discharge at which bank full flow occurs. As a result of urbanization, the frequency of discharges which fill the channel will increase, with the effect that the beds and banks of channels in credible materials will be eroded so as to enlarge the channel. This in turn will lead to bank caving, possible undermining of structures, and increases in turbidity (Hollis and Luckett, 1976).

Comparable changes in channel morphology result from discharge diminution produced by flood-control works and diversions for irrigation. This can be shown for the North Platte and the South Platte in America, where both peak discharge and mean annual discharge have declined to 10–30 per cent of their pre-dam values. The North Platte, 762–1219 m wide in 1890 near the Wyoming-Nebraska border, has narrowed to about 60 m at present; while the South Platte River was about 792 m wide, 89 km above its junction with the North Platte in 1897, but had narrowed to about 60 m by 1959 (Schumm, 1977: 161). The tendency of both rivers has been to form one narrow, well-defined channel in place of the previously wide, braided channels, and, in addition, the new channel is generally somewhat more sinuous than the old.

Similarly, the building of dams can lead to channel aggradation upstream from the reservoir and channel deepening downstream because of the changes brought about in sediment loads. Some data on observed rates of degradation below dams are presented in Table 1.11. They show that the average rate of degradation has been of the order of a few meters over a few decades following closure of the dams. However, over time the rate of degradation seems to become less or to cease altogether, and Leopold *et al.* (1964: 455) suggest that this can be brought about in several ways. First, because degradation results in a flattening of the channel slope in the vicinity of the dam — the slope may become so flat that the necessary force to transport the available materials is no longer provided by the flow. Second, the reduction of flood peaks by the dam reduces the competence of the transporting stream to carry some of the material on its bed. Thus if the bed contains a mixture of particle size the river may be able to transport the finer sizes but not the larger, and the gradual winnowing of the fine particles will leave an armour of coarser material that prevents further degradation.

Table 1.11 River-bed degradation below dams

River	*Dam*	*Amount (m)*	*Length (km)*	*Time (years)*
South Canadian (USA)	Conchos	3.1	30	10
Middle Loup (USA)	Milburn	2.3	8	11
Colorado (USA)	Hoover	7.1	111	14
Colorado (USA)	Davis	6.1	52	30
Red (USA)	Denison	2.0	28	3
Cheyenne (USA)	Angostura	1.5	8	16
Saalach (Austria)	Reichenhall	3.1	9	21
South Saskatchewan (Canada)	Diefenbaker	2.4	8	12
Yellow (China)	Samenxia	4.0	68	4

Source: from data in Galay, 1983.

The overall effect of the creation of a reservoir by the construction of a dam is to lead to a reduction in downstream channel capacity (see Petts, 1979 for a review). This seems to amount to between about 30 and 70 per cent (see Table 1.12).

Table 1.12 Channel capacity reduction below reservoirs

River	*Dam*	*% channel capacity loss*
Republican, USA	Harlon County	66
Arkansas, USA	John Martin	50
Rio Grande, USA	Elephant Buttre	50
Tone, UK	Clatworthy	54
Meavy, UK	Burrator	73
Nidd, UK	Angram	60
Burn, UK	Burn	34
Derwent, UK	Ladybower	40

Source: modified after Pells, 1979.

Equally far-reaching changes in channel form are produced by land-use changes and the introduction of soil conservation measures. Clearing of the land for cultivation caused massive slope erosion which resulted in the transference of large quantities of sediment into channels and flood-plains. This phase of intense erosive land-use persisted and was particularly strong during the nineteenth century and the first decades of the twentieth century, but thereafter conservation measures, reservoir construction and a reduction in the intensity of agricultural land use led to further channel changes (Trimble, 1974). Streams ceased to carry such a heavy sediment load, they became much less turbid, and incision took place into the flood-plain sediments. By means of this active stream-bed erosion, streams incised themselves into the modern alluvium, lowering their beds by as much as 3–4 m.

In the Platte catchment of South-West Winsconsin a broadly comparable picture of channel change has been documented by Knox (1977). There, as in the Upper Mississippi valley (Knox, 1987), it is possible to identify stages of channel modification associated with various stages of land use, culminating in decreased overbank sedimentation as a better of ratter land management in the last half century.

Other significant changes produced in channels include those prompted by accelerated sedimentation associated with changes in the vegetation communities growing along channels. The introduction of the salt cedar in the Southern USA has caused significant flood-plain aggradation. In the case of the Brazos River in Texas, for example, the plants encouraged sedimentation by their damming and ponding effect. They clogged channels by invading sand banks and sand bars, and so increased the area subject to flooding. Between 1941 and 1979 the channel width declined from 157 to 67 m, and the amount of aggradation was as much as 5.5 m (Blackburn *et al.*, 1983). Another example of rapid aggradation is that produced by the introduction of mining

Table 1.13 Causes of river-bed degradation

Type	*Primary cause*	*Contributory cause*
Downstream progressing	Decrease of bed-material discharge	Dam construction Excavation of bed material Diversion of bed material Change in land use Storage of bed material
	Increased water discharge	Diversion of flow Rare floods
	Decrease in bed-material size	
	Other	River emerging from lake Thawing of permafrost
Upstream progressing	Lower base level	Drop in lake level Drop in level of main river Excavation of bed material
	Decrease in river length	Cut-offs Channelization Stream capture Horizontal shift of base level
	Removal of control point	Natural erosion Removal of dam

Source: after Galay, 1983.

waste into channels. In mid-Wales the Afon Ystwyth suffered 3 m of aggradation for this reason (Lewin *et al.*, 1983).

Overall, the causes of observed cases of river-bed degradation are varied and complex and result from a variety of natural and human changes (Table 1.13). A useful distinction can be drawn between degradation that proceeds downstream, and that which proceeds upstream, but in both cases the complexity of causes is evident.

Reactivation and stabilization of sand dunes

To George Perkins Marsh the reactivation and stabilization of sand dunes, especially coastal dunes, was a theme of the greatest importance in his analysis of human transformation of nature. He devoted fifty-four pages to it:

> The preliminary steps, whereby wastes of loose, drifting, barren sands are transformed into wooded knolls and plains, and finally through the accumulation of vegetable mould, into arable ground, constitute a conquest over nature which precedes agriculture— a geographical revolution—and therefore, an account of the means by which the change has been effected belongs properly to the history of man's influence on the great features of physical geography. (1965: 393).

He was fascinated by 'the warfare man wages with the sand hills' and asked (1965: 410) 'in what degree the naked condition of most dunes is to be ascribed to the improvidence and indiscretion of man'.

His analysis showed quite clearly that most of the coastal dunes of Europe and North America had been rendered mobile, and hence a threat to agriculture and settlement, through human action, especially because of grazing and clearing. In Britain the cropping of dune warrens by rabbits was a severe problem, and a most significant event in their long history was the myxomatosis outbreak of the 1950s, which severely reduced the rabbit population and led to dramatic changes in stability and vegetative cover

Appreciation of the problem of dune reactivation on mid-latitude shorelines, and attempts to overcome it, go back a long way (Kittredge, 1948). For example, the menace of shifting sand following denudation is recognized in a decree of 1539 in

Denmark which imposed a fine on those who destroyed certain species of sand plants on the coast of Jutland. The fixation of coastal sand dunes by planting vegetation was initiated in Japan in the seventeenth century, while attempts at the reafforestation of the spectacular Landes dunes in South-West France began as early as 1717, and came to fruition in the nineteenth century through the plans of the great Bremontier: 81,000 hectarės of moving sand had been fixed in the Landes by 1865. In Britain possibly the most impressive example of sand control is provided by the reafforestation of the Culbin Sands in North-East Scotland with conifer plantations (Edlin, 1976).

Human-induced dune instability is not, however, a problem that is restricted to mid-latitude coasts. In inland areas of Europe, clearing, fire and grazing have affected some of the late Pleistocene dune fields that were created on the arid steppe margins of the great ice sheets, and in Eastern England the dunes of the Breckland presented problems on many occasions. There are records of carriages being halted by sand-blocked roads and of one village, Downham, being overwhelmed altogether.

However, it is possibly on the margins of the great subtropical and tropical deserts that some of the strongest fears are being expressed about sand-dune reactivation. This is one of the facets of the process of desertification and desertization. The increasing population levels of both humans and their domestic animals, brought about by improvements in health and by the provision of boreholes, has led to an excessive pressure on the limited vegetation resources. As ground cover has been reduced, so dune instability has increased. The problem is not so much that dunes in the desert cores are relentlessly marching on to moister areas, but that fossil dunes, laid down during the more arid phase peaking around 18,000 years ago, have been reactivated *in situ.*

A wide range of methods is available to attempt to control drifting sand and moving dunes as follows:

1. *Drifting sand*

   Enhancement of deposition of sand by the creation of large ditches, vegetation belts and barriers and fences

   Enhancement of transport of sand by aerodynamic

streamlining of the surface, change of surface materials or panelling to direct flow

Reduction of sand supply, by surface treatment, improved vegetation cover or erection of fences

Deflection of moving sand, by fences, barriers, tree belts, etc.

2. *Moving dunes*

Removal by mechanical excavation

Destruction by reshaping, trenching through dune axis or surface stabilization of barchan arms

Immobilization, by trimming, surface treatment and fences.

Yet in practice most solutions to the problem of dune instability and sand blowing have involved the establishment of a vegetation cover. This is not always easy. Species used to control sand dunes must be able to endure undermining of their roots, burning, abrasion and often severe deficiencies of soil moisture. Thus the species selected need to have the ability to recover after partial burying, to have deep and spreading roots, to have rapid height growth in the seedling stages, to promote rapid litter development, and to add nitrogen to the soil through root nodules. During the early stages of growth they may need to be protected by fences, sand traps and surface mulches. Growth can also be stimulated by the addition of synthetic fertilizers.

In the hearts of deserts sand dunes are naturally mobile because of the sparse vegetation cover. Even here, however, humans sometimes attempt to stabilize sand surfaces to protect settlements, pipelines, industrial plant and agricultural land. The use of relatively porous barriers to prevent or divert sand movement has proved comparatively successful, and palm fronds or chicken wire have made adequate stabilizers. Elsewhere surfaces have been strengthened by the application of high-gravity oil or by salt-saturated water (which promotes the development of wind-resistant surface crusts).

In temperate areas coastal dunes have been effectively stabilized by the use of various trees and other plants (Ranwell and Boar, 1986). In Japan *Pinus thümbergii* has been successful, while in the great Culbin Sands plantations of Scotland *P. nigra* and

P. laricio have been used initially, followed by P. *sylvestris*. Of the smaller shrubs, *Hippophae* has proved highly efficient, sometimes too efficient, at spreading. Its clearance from areas where it is not welcome is difficult precisely because of some of the properties that make it such an efficient sand stabilizer: vigorous suckering growth and the rapid regrowth of cut stems (Boorman, 1977). Different types of grass have also been employed, especially in the early stages of stabilization. These include two grasses which are moderately tolerant of salt: *Agrophyron junceiforme*, now called *Elymus farctus* (sand twitch) and *Lye arenarius,* now called *Leymus arenarius* (lyme grass). Another grass which is much used, not least because of its rapid and favourable response to burial by drifting sand, is *Ammophila arenaria* (marram).

Further stabilization of coastal dunes has been achieved by setting up sand fences. These generally consist of slats about 1.0–1.5 m high, and have a porosity of 25–50 per cent. They have proved to be effective in building incipient dunes in most coastal areas. By installing new fences regularly, large dunes can be created with some rapidity. Alternative methods, such as using junk cars on the beaches at Galveston, Texas, have been attempted with little success.

Accelerated coastal erosion

Because of the high concentration of settlements, industries, transport facilities and recreational developments on coastlines, the pressures placed on coastal landforms are often acute, and the consequences of excessive erosion serious. While most areas are subject to some degree of natural erosion and accretion, the balance can be upset by human activity in a whole range of different ways (Table 1.14). However, humans seldom attempt to accelerate coastal erosion deliberately. More usually, it is an unexpected or unwelcome result of various economic projects. Frequently coast erosion has been accelerated as a result of human efforts to reduce it.

One of the best forms of coastal protection is a good beach. If material is removed from a beach accelerated cliff retreat may take place. Removal of beach materials may be necessary to secure valuable minerals, including heavy minerals, or to provide aggregates for construction. The classic example of the latter was

Table 1.14 Mechanisms of human-induced erosion in coastal zones

Human-induced erosion zones	*Effects*
Beach mining for placer deposits (heavy minerals) such as zircon, rutile, ilmenite and monazite	Loss of sand from frontal dunes and beach ridges
Construction of groynes, breakwaters, jetties and other structures	Downdrift erosion
Construction of offshore breakwaters	Reduction in littoral drift
Construction of retaining walls to maintain river entrances	Interruption of littoral drift resulting in downdrift erosion
Construction of seawalls, revetments, etc.	Wave reflection and accelerated sediment movement
Deforestation	Removal of sand by wind: sand drift
Fires	Migrating dunes and sand drift after destruction of vegetation
Grazing of sheep and cattle	Initiation of blow-outs and transgressive dunes: sand drift
Off-road recreational vehicles (dune buggies, trail bikes, etc.)	Triggering mechanism for sand drift attendant upon removal of vegetative cover
Reclamation schemes	Changes in coastal configuration and interruption of natural processes, often causing new patterns in sediment transport
Increased recreational needs	Accelerated deterioration, and destruction, of vegetation on dunal areas, promoting erosion by wind and wave action

Source: Hails, 1977.

the mining of 6,60,000 tonnes of shingle from the beach at Hallsands in Devon, England, in 1887 to provide material for the construction of dockyards at Plymouth. The shingle proved to be undergoing little or no natural replenishment and in consequence the shore level was reduced by about 4 m. The loss of the protective shingle soon resulted in cliff erosion to the extent of 6 m between 1907 and 1957. The village of Hallsands was cruelly attacked by waves and is now in ruins.

Another common cause of beach and cliff erosion at one point is coast protection at another. As already stated, a broad beach serves to protect the cliffs behind, and beach formation is often encouraged by the construction of groynes. However, these structures sometimes merely displace the erosion (possibly in an even more marked form) further along the coast.

Piers or breakwaters can have similar effects to groynes. This has occurred at various places along the British coast: erosion at Seaford resulted from che Newhaven breakwater, while erosion at Lowestoft resulted from the pier at Gorleston. At Madras in South-East India, for example, a 1000 m long breakwater was constructed in 1875 to create a sheltered harbour on a notoriously inhospitable coast, dominated by sand transport from north to south. On the south side of the breakwater over 1 million m^2 of new land formed by 1912, but erosion occurred for 5 km north of the breakwater. At Ceara in Brazil, also in 1875, a detached breakwater was erected over a length of 430 m more or less parallel to the shore. It was believed that by using a detached structure littoral drift would be able to move along the coast uninterrupted by the presence of a conventional structure built across the surf zone. This, however, proved to be a fallacy (Komar, 1976), since the removal of the wave action which provided the energy for transporting the littoral sands resulted in their deposition within the protected area.

As time goes on the beach in the foreground builds outwards, while the cliff behind the jetty retreats and needs to be protected with a seawall. Likewise, the construction of some seawalls, erected to reduce coastal erosion and flooding, has had the opposite effect to the one intended. Given the extent to which artificial structures have spread along the coastlines of the world, this is a serious matter (Walker, 1988).

Problems of this type are exacerbated because there is now abundant evidence to suggest that much of the reservoir of sand and shingle that creates beaches is in some respects a relict feature. Much of it was deposited on the continental shelf during the maximum of the last glaciation (around 18,000 year BP), when sealevel was about 120–40 m below its present level. It was transported shoreward and incorporated in present-day beaches during the phase of rapidly rising post-Glacial sealevels that characterized the Flandrian transgression until about 6000 years BP. Since that time, with the exception of minor oscillations of the order of a few metres, world sealevels have been stable and much less material is, as a consequence, being added to beaches and shingle complexes. Therefore, according to Hails (1977: 322): 'in many areas, there is virtually no offshore supply to be moved onshore, except for small quantities resulting from seasonal changes.' It is because of these problems that many erosion prevention schemes now involve beach replenishment (by the artificial addition of appropriate sediments to build up the beach), or employ miscellaneous sand bypassing techniques (including pumping and dredging) whereby sediments are transferred from the accumulation side of an artificial barrier to the erosional side (King, 1974).

In some areas, however, sediment-laden rivers bring material into the coastal zone which becomes incorporated into beaches through the mechanism of longshore drift. Thus any change in the sediment load of such rivers may result in a change in the sediment budget of neighbouring beaches. When accelerated soil erosion occurs in a river basin the increased sediment load may cause coastal accretion and siltation. But where the sediment load is reduced through action such as the construction of large reservoirs, behind which sediments accumulate, coastal erosion may result. This is believed to be one of the less desirable consequences of the construction of the Aswan Dam on the Nile: parts of its delta have shown recently accelerated recession.

The Nile sediments, on reaching the sea, used to move eastward with the general anticlockwise direction of water movement in that part of the Eastern Mediterranean, generating sand bars and dunes which contributed to delta accretion. About a century ago an inverse process was initiated and the delta began to retreat. For example, the Rosetta mouth of the Nile has lost about 1.6 km of its length from 1898–1954. The imbalance between sedimentation and erosion appears to have started with the Delta Barrages (1861) and then been continued by later works, including the Sennar Dam (1925), Gebel Aulia Dam (1937), Khasm el Girba Dam (1966), Roseires Dam (1966) and the High Dam itself. Much of the Egyptian coast is now 'undernourished' with sediment and, as a result of this overall erosion of the shoreline, the sand bars bordering Lake Manzala and Lake Burullus on the seaward side are eroded and likely to collapse. If this were to happen, the lakes would be converted into marine bays, so that saline water would come into direct contact with low-lying cultivated land and freshwater aquifers.

Likewise in Texas, where over the last century four times as much coastal land has been lost as has been gained, one of the main reasons for this change is believed to be the reduction in the suspended loads of some of the rivers discharging into the Gulf of Mexico (table 1.15). The four rivers listed carried, in 1961–70, on average only about one-fifth of what they carried in 1931–40. Comparably marked falls in sediment loadings occurred elsewhere in the Eastern USA. Likewise, in France the once mighty Rhône only carries about 5 per cent of the load it did in the nineteenth century; and in Asia, the Indus discharges less than 20 per cent of the load it did before construction of large barrages over the last half century (Milliman, 1990).

Table 1.15 Suspended loads of Texas rivers discharging into the Gulf of Mexico

River	*Suspended load (million tonnes)* 1331–40	1961–70	%*
Brazos	350	120	30
San Bernard	1	1	100
Colorado	100	11	10
Rio Grande	180	6	3
Total	631	138	20

*1961–70 loads as % 1931–40 loads

Source: modified from Hails, 1977.

Construction of great levees on the Mississippi River since 1717 has also affected the Gulf of Mexico coast. The channelization of the river has increased its velocity, reduced overbank deposition of silt onto swamps, marshes and estuaries, and changed the salinity conditions of marshland plants (Cronin, 1967). As a result, the coastal marshes and islands have suffered from increased erosion or a reduced rate of development. This has been vividly described by Biglane and Lafleur (1967: 691):

> Like a bullet through a rifle barrel, waters of the mighty Mississippi are thrust toward the Gulf between the confines of the flood control levees. Before the day of these man-made structures, these waters poured out over tremendous reaches of the coast.... Freshwater marshes (salinities averaging 4–6%) were formed by deposited silts and vegetative covers of wire grass.... As man erected his flood protection devices, these marshes ceased to form as extensively as before.

However, as with so many examples of environmental change, it is unlikely that just one factor, in this case channelization, is the sole cause of the observed trend. In their study of erosion loss in the Mississippi Delta and neighbouring parts of the Louisiana coast, Walker *et al.* (1987) suggest that this loss is the result of a variety of complex interactions among a number of physical, chemical, biological and cultural processes. These processes include, in addition to channelization, worldwide sea-level changes, subsidence resulting from sediment loading by the delta of the underlying crust, changes in the sites of deltaic sedimentation as the delta evolves, catastrophic storm surges and subsidence resulting from subsurface fluid withdrawal.

In some areas anthropogenic vegetation modification creates increased erosion potential. This has been illustrated for the hurricane-afflicted coast of Belize, Central America (Stoddart, 1971). He showed that natural, dense vegetation thickets on low, sandy islands *(cays)* acted as a baffle against waves and served as a massive sediment trap for coral blocks, shingle and sand transported during extreme storms. However, on many islands the natural vegetation had been replaced by coconut plantations. These had an open structure easily penetrated by sea water, they tended to have little or no ground vegetation (thus exposing the cay surface to stripping and channelling), and they had a dense but shallow root net easily undermined by marginal sapping. Thus Stoddart found that 'where the natural vegetation had been replaced by coconuts before the storm (Hurricane Hattie), erosion and beach retreat led to net vertical decreases in height of 3–7 ft; whereas where natural vegetation remained, banking of storm sediments against the vegetation hedge led to a net vertical increase in height of 1–5 ft.'

Other examples of markedly accelerated coastal erosion and flooding result from anthropogenic degradation of dune ridges. Frontal dunes are a natural defence against erosion, and coastal changes may be long-lasting once they are breached. Many of those areas in Eastern England which most effectively resisted the great storm and surge of 1952 were those where humans had not intervened to weaken the costal dune belt.

Not all dune stabilization and creation schemes have proved desirable (Dolan *et al.*, 1973). In North Carolina the natural barrier-island system along the coastline met the challenge of periodic extreme storms, such as hurricanes, by placing no permanent obstruction in the path of the powerful waves. Under these natural conditions, most of the initial stress of such storms is sustained by relatively broad beaches. Since there is no resistance from impenetrable landforms, water can flow between the dunes (which do not form a continuous line) and across the islands, with the result that wave energy is rapidly exhausted. However, between 1936 and 1940, 1000 km of sand fencing was erected to create an artificial barrier dune along part of the Outer Banks, and 2.5 million trees and various grasses (especially *Ammophila breviligulata)* were planted to create large artificial dunes. The altered barrier islands not only have the artificial barrier-dune system; they also have beaches that are often only 30 m wide, compared with 140 m for the unaltered islands. This beach-narrowing process, combined with the presence of a permanent dune structure, has created a situation in which high wave energy is concentrated in an increasingly restricted run up area, resulting in a steeper beach profile, increased turbulence and greater erosion. Another problem associated with artificial dune stabilization is the flooding that occurs when north-east storms pile the water of the lagoon, Pamlico Sound, up against the barrier island. In the past, these surge waters simply flowed out between the low, discontinuous dunes and over the beach to the sea, but with the altered dune chain the water cannot drain off readily and vast areas of land are at times submerged.

The human impact on seismicity and volcanoes

The seismic and tectonic forces which mould the relief of the earth and cause such hazards to human civilization are one of the fields in which efforts to control natural events have had least success and where least has been attempted. Nonetheless, the fact that humans have been able, inadvertently, to trigger off small earthquakes by nuclear blasts, as in Nevada (Pakiser *et al.*, 1969), by injecting water into deep wells, as in Colorado, by mining, by building reservoirs, and by fluid extraction, suggests that in due course it may be possible to 'defuse' earthquakes by relieving tectonic strains gradually in a series of non-destructive, low-intensity earthquakes. One problem, however, is that there is no assurance that an earthquake purposefully triggered by human action will be a small one, or that it will be restricted to a small area. The legal implications are immense.

The demonstration that increasing water pressures could initiate small-scale faulting and seismic activity was unintentionally demonstrated near Denver (Evans, 1966), where nerve-gas waste was being disposed of at great depth in a well in the hope of avoiding contamination of useful groundwater supplies. The waste was pumped in at high pressures and triggered off a series of earthquakes, the timing of which corresponded very closely to the timing of waste disposal in the well. It is also now thought that the pumping of fluids into the Inglewood Oil Field, Los Angeles, to raise the hydrostatic pressure and increase oil recovery, may have been responsible for triggering the 1963 earthquake which fractured a wall of the Baldwin Hills Reservoir. It appears that increased fluid pressure reduces the frictional force across the contact surface of a fault, allows slippage to occur, thereby causing an earthquake.

In general earthquake triggering has been related to fluid injection, but for reasons that are still obscure there may be some cases where increased seismicity has resulted from fluid abstraction (Segall, 1989).

The significance of these 'accidents' has now been verified experimentally at an oilfield at Rangley in Colorado, USA, where variations in seismicity have been produced by deliberately controlled variations in the fluid pressure in a zone that is seismically active (Raleigh *et al.*, 1976).

Perhaps the most important anthropogenically induced seismicity results from the creation of large reservoirs. Reservoirs impose stresses of significant magnitude on crustal rocks at depths rarely equalled by any other human construction. With the ever-increasing number and size of reservoirs the threat rises. There are at least six cases (Koyna, Kremasta, Hsinfengkiang, Kariba, Hoover and Marathon) where earthquakes of a magnitude greater than 5, accompanied by a long series of foreshocks and aftershocks, have been related to reservoir impounding. However, as Table 1.16 shows, there are many more locations where the filling of reservoirs behind dams has led to appreciable, though less dramatic, levels of seismic activity. Detailed monitoring has shown that earthquake clusters occur in the vicinity of some dams after their reservoirs have been filled, whereas before construction activity was less clustered and less frequent. Similarly, there is evidence from Vaiont (Italy), Lake Mead (USA), Kariba (Central Africa), Koyna (India) and Kremasta (Greece) that there is a linear correlation between the storage level in the reservoir and the logarithm of the frequency of shocks. This is illustrated for Vaiont, Koyna and Nurek. It is also apparent from Nurek that as the great reservoir has filled, so the depth of the more shallow-seated earthquakes appears to have increased.

One reason why dams induce earthquakes involves the hydroisostatic pressure exerted by the mass of the water impounded in the reservoir, together with changing water pressures across the contact surfaces of faults. Given that the deepest reservoirs provide surface loads of only 20 bars or so, direct activation by the mass of the impounded water seems an unlikely cause (Bell and Nur, 1978) and the role of changing pore pressure assumes greater importance. Paradoxically, there are some possible examples of reduced seismic activity induced by reservoirs. (Milne, 1976). One possible explanation of this is the increased incidence of stable sliding (fault creep) brought about by higher pore-water pressure in the vicinity of the reservoir.

However, the ability to prove an absolutely concrete cause-and-effect relationship between reservoir activity and

Table 1.16 List of dams for which associated seismic phenomena are reported

Dam	*Year of completion*	*Year of largest earthquake since dam completion*
Marathon, Greece	1929/30	1938
Oued Fodda, Algeria	1932	–
Boulder, USA	1936	1939
Pieve di Cadore, Italy	1949	–
Clark Hill, USA	1952	1974
Kariba, Zimbabwe/Zambia	1959	1963
Grandval, France	1959	1963
Hsingfengkiang, China	1959	1962
Kurobe, Japan	1960	1961
Camarillas, Spain	1960	1961
Cannelles, Spain	1960	1962
Vaiont, Italy	1961	1963
Monteynaid, France	1962	1963
Koyna, India	1962	1967
Benmore, New Zealand	1965	1966
Kremasta, Greece	1965	1966
Piastra, Italy	1965	1966
Contra, Switzerland	1965	1965
Najina Basta. Yugoslavia	1966	1967
Grancarevo, Yugoslavia	1967	–
Oroville, USA	1968	1975
Kastraki, Greece	1968	–
Nurek, Tajik Republic	1969	1972
Vouglans, France	1970	1971
Karnafusa, Japan	1970	–
Talbingo, Australia	1971	1972
Schlegeis, Austria	1971	–
H. Verwoerd, South Africa	1972	–
Jocasses, USA	1972	1975
Keban, Turkey	1973	1974
Manic 3, Canada	1975	1975

Source: from data in Judd, 1974 and Milne, 1976, processed by author.

earthquakes is severely limited by our inability to measure stress below depths of several kilometres, and some examples of induced seismicity may have been built on the false assumption that because an earthquake occurs in proximity to a reservoir it has to be induced by that reservoir (Meade, 1991).

Miscellaneous other human activities appear to affect seismic levels. In Johannesburg, South Africa, for example, gold-mining and associated blasting activity have produced ten of thousands of small tremors, and there is a notable reduction in the number that occur on Sundays, a day of rest. In Staffordshire, England, coal-mining has caused increased seismic activity. There are also cases where seismicity and faulting can be attributed to fluid extraction, for example, in the oilfields of Texas and California and the gasfields of the Po Valley in Italy.

When looking at the human impact on volcanic activity human impotence becomes apparent, though some success has been achieved in the control of lava flows. Thus in 1937 and 1947 the US Army attempted to divert lava from the city of Hilo, Hawaii, by bombing threatening flows, while elsewhere, where lava rises in a crater, breaching of the crater wall to direct lava towards uninhabited ground may be possible. In 1973 an attempt was made to halt advance of lava with cold water during the Icelandic eruption of Kirkjufell. Using up to 4 million litres of pumped water per hour, the lava was cooled sufficiently to decrease its velocity at the flow front so that the chilled front acted as a dam to divert the still fluid lava behind (Williams and Moore, 1973).

❑❑❑

2

The Human Impact on Climate and the Atmosphere

World climates

THE CLIMATE OF THE WORLD is now known to have fluctuated frequently and extensively in the three or so million years during which humans have inhabited the earth (Goudie, 1992). The bulk of these changes have nothing to do with human intervention. Climate has changed, and is currently changing, because of a wide range of different natural factors which operate over a variety of time scales.

Nonetheless, with the increasing human population and the rising level of technology, it is now apparent that over the last century human agency has probably become a significant factor in the variations in world climate which are taking place, but it remains extremely difficult to ascertain whether human influence or natural forces are responsible for observed trends. If human activity can be considered a factor then this is particularly because of inadvertent effects on atmospheric quality and on the albedo of land masses. Human influence on global climate can be summarized as follows:

Possible mechanisms

Gas emissions

CO_2 – industrial and agricultural

methane

chlorofluorocarbons (CFCs)
nitrous oxide
krypton 85
water vapour
miscellaneous trace gases

Aerosol generation
Thermal pollution
Albedo change
dust addition to ice caps
deforestation
overgrazing

Extension of irrigation
Alteration of ocean currents by constricting straits
Diversion of fresh waters into oceans

The CO_2 problem

Since the beginning of the Industrial Revolution humans have been taking stored carbon out of the earth in the form of coal, petroleum and natural gas, and burning it to make carbon dioxide (CO_2), heat, water vapour and smaller amounts of sulphur dioxide (SO_2) and other gases.

The pre-industrial level of carbon dioxide is a matter of some debate, but may have been as low as 260–70 ppm by volume (Wigley, 1983). The present level is over 350 ppmv, and the upward trend is evident in records from various parts of the world. Forecasts suggest that the level by 2065 will be about 600 ppm (Carbon Dioxide Assessment Committee, 1983). If fossil-fuel consumption continues at present rates, a concentration of up to 2000 ppm of CO_2 might be reached in two or three centuries from now. However, predictions are not without their problems, depending as they do on two major imponderables: the future rate of fossil-fuel consumption, and the way in which CO_2 may be absorbed by the oceans.

The prime cause of the release of CO_2 into the atmosphere is probably the burning of fossil fuels. Fossil-fuel burning throughout the world in 1981 released about 5.3 gigatons of carbon to the atmosphere as CO_2. The cumulative emissions from the mid-ninteenth century to 1981 totalled about 160 gigatons of carbon as CO_2. Fossil-fuel use for much of the post-Second World War period averaged an increase of around 4.5 per cent per year, but because of the changing nature of energy provision resulting from the massive increases in the cost of oil, and other factors such as the adoption of nuclear power, it is now widely believed that energy consumption using fossil fuels will rise relatively more slowly in the near future, probably averaging about 2.5 per cent per year.

Another factor that may contribute substantially to CO_2 levels in the atmosphere is the burning of forests and changes in the organic levels in soils that are subjected to deforestation and cultivation (Wilson, 1978; Wong, 1978). Woodwell (1978) estimated that this source may be currently more important than the burning of fossil fuels, with deforestation releasing about 6 gigatons of carbon as CO_2 per year, and the decay of soil humus about 2 gigatons of carbon as carbon dioxide per year. However, there is considerable uncertainty about existing rates of deforestation, the proportion of deforestation that is caused by burning, the amount of carbon that may be kept stored as charcoal, and the rate at which regrowth of vegetation consumes carbon. Consequently, the rates initially proposed by Woodwell have now been reassessed, and a current consensus value for the amount of carbon dioxide released by land-use changes is of the order of 2 gigatons per annum (Clark, 1982). Attempts have also been made to assess longer-term release of carbon dioxide into the atmosphere by undertaking studies either of historical changes in land use, or by analysing $^{13}C/^{12}C$ ratios in tree rings. Both methods have given broadly similar results and indicate that since the middle of the last century human activities have released on average between 0.7 and 1.7 gigatons per year. From 1860 to the present, cumulative releases from changes in land use have been approximately equal to cumulative releases from the burning of fossil fuels.

The CO_2 levels in the atmosphere have an effect on the global heat balance, since CO_2 is virtually transparent to incoming solar radiation but absorbs outgoing terrestrial infrared radiation; radiation that would otherwise escape to space and result in loss of heat from the lower atmosphere. This is called 'the greenhouse effect.'

Consequently one would expect increased CO_2 levels to lead to an increase in surface temperatures (Hansen *et al.*, 1981). Indeed, there is some evidence of an overall global warming trend of about 0.5°C in this century (see, for example, Jones *et al.*, 1986) and the 1980s saw an exceptionally warm decade.

In reality the term 'greenhouse effect' is something of a misnomer. As Henderson-Sellers and Robinson (1986: 60) explain:

> We now know that a greenhouse maintains its higher internal temperature largely because the shelter it offers reduces the turbulent transfers of energy away from the surface rather than because of any radiative considerations. Thus while the greenhouse effect remains valid, and vital, for the atmosphere it might be better to think of the physical processes in terms of the 'leaky bucket' analogy.... Here an increase m the amount of gas with absorption bands in the infrared part of the spectrum is represented by a decrease in the size of the hole in the bucket. The surface temperature, represented by the depth of the water in the bucket, rises as more absorbing gases enter the atmosphere.

Other gases

In addition to carbon dioxide, it is probable that other gases will contribute to the greenhouse effect. Individually their effects may be minor, but as a group they may be major (Ramanathan, 1988). Indeed, molecule for molecule some of them may be much more effective as greenhouse gases than CO_2, as the data in Table 2.1 show.

Table 2.1 Radiative forcing relative to CO_2 per unit molecule change in the atmosphere

Gas	*Relative radiative forcing*
CO_2	1
CH_4 (methane)	21
N_2O (nitrous oxide)	206
CFC-11	12,400
CFC-12	15,800

Source: extracted from Houghton *et al.*, 1990, p. 53.

One of the more important of the trace gases is methane, which has a strong infrared absorption band at 7.66 μm. Ice core studies and recent direct observations suggest that until the beginning of the eighteenth century background levels were stable at around 600 parts per billion by volume (ppbv). They rose steadily between AD 1700 and 1900, and then increased still more rapidly, attaining levels that averaged 1300 ppbv in the early 1950s and 1600 ppbv by the mid-1980s (Khalil and Rasmussen, 1987). This increase of 2.5 times over background levels results primarily from increased rice cultivation in waterlogged paddy fields, the enteric fermentation produced in the growing numbers of flatulent domestic cattle, and the burning of oil and natural gas (Crutzen *et al.*, 1986) (Table 2.2).

Table 2.2 (a) Estimates of methane (CH_4) source strengths and sinks

Sources/sinks	*Best estimate ($10^6t\ a^{-1}$)*	*Range ($10^6t\ a^{-1}$)*
Sources		
Natural wetlands	115	100–200
Rice paddies	110	25–170
Enteric fermentation (animals)	80	65–100
Gas drilling, venting, transmission	45	25–50
Biomass burning	40	20–80
Termites	40	10–100
Landfills	40	20–70
Coal mining	35	19–50
Oceans	10	5–20
Fresh waters	5	1–25
CH_4 hydrate destabilization	5	0–100
Sinks		
Removal by soils	30	15–45
Reaction with OH	500	400–600
Atmospheric increase	44	40–48

(b) Estimates of nitrous oxide (N_2O) source strengths and sinks

Sources/sinks	*Range* ($10^6 t\ a^{-1}$)
Sources	
Oceans	1.4–2.6
Soils (tropical forests)	2.2–3.7
Soils (temperate forests)	0.7–1.5
Fossil fuel combustion	0.1–0.3
Biomass burning	0.02–0.2
Fertilizer (including groundwater)	0.01–2.2
Sinks	
Removal by soils	Unknown
Photolysis in the stratosphere	7–13
Atmospheric increase	3–4.5

Source: data in UNEP, 1991.

Chlorofluorocarbons (CFCs), despite their relatively trace amounts in the atmosphere, have increased very markedly in terms of their emissions and their concentrations in recent decades, and have a very strong greenhouse effect even in relatively small amounts. On the other hand, the ozone depletion they have caused in the stratosphere may counteract this effect, for stratospheric ozone depletion results in a decrease in radiative forcing (Houghton *et al*., 1992).

Nitrous oxide (N_2O) is also no laughing matter, for it can contribute to the greenhouse effect, primarily by absorption of infrared at the 7.8 and 17 μm bands. Combustion of hydrocarbon fuels, the use of ammonia-based fertilizers, deforestation and biomass burning are among the processes that could lead to increase in atmospheric N_2O levels.

Warnings have been circulated (Boeck *et al*., 1975) about the possible role of a substance emitted from nuclear reactors, called Krypton 85. It is believed that its increasing presence will reduce the electrical resistance of the atmosphere between the oceans and the ionosphere. This in turn would affect the electrification of thunderclouds and, through that, precipitation levels.

Supersonic aircraft, both civil and military, discharge some water vapour into the stratosphere as contrails. At present, the water content of the stratosphere is low, as is the exchange of air between the lower stratosphere and other regions. Consequently, comparatively modest amounts of water vapour discharged by aircraft could have a significant effect on the natural balance. It has been calculated that 400 supersonic aircraft, making four flights per day, would place 150 million kg of water into the lower stratosphere. This could, over a period of years, double the existing water content of the atmosphere, thereby leading to a small rise in temperature of perhaps 0.6°C (Sawyer, 1971).

However, statistics on the increase of cirrus levels as a result of stratospheric aircraft movements are still meagre (SCEP Report, in Matthews *et al.*, 1971: 39) and their statistical significance is not always proven. Yet studies at Salt Lake City and Denver between 1949 and 1969 do show an upward trend in cirrus cloudiness which paralleled the increase in jet aircraft activity. The expected rapid development of civil aviation in future decades may make this an important element of anthropogenic atmospheric modification.

Other trace gases that could play a greenhouse role include bromide compounds, carbon tetrafluoride, carbon tetrachloride and methyl chloride.

It is possible that the increase in global surface temperatures to be expected by 2050 as a result of these trace gases is 1.4–2.2°C, an increase that may be comparable to that anticipated from CO_2 itself (Macdonald, 1982). Indeed, Bolin *et al.*, 1986: xxi) suggest:

> The role of greenhouse gases other than CO_2 in changing the climate is already about as important as that of CO_2. If present trends continue, the combined concentrations of atmospheric CO_2 and other greenhouse gases would he radiatively equivalent to a doubling of CO_2 from pre-industrial levels possibly as early as the 2030s.

Aerosols

One consequence of the Industrial Revolution, which may be of equal significance to CO_2 emission, is the increased quantity of dust or smoke particles that are emitted to the lower atmosphere.

A striking illustration of this is provided by the analysis of dust levels in glacier ice of known age in the Southern CIS. Layers of ice dated AD 1800–1920 show a dust content of 10 mg/l, whereas by the 1950s, the figure had increased twenty rimes to 200 mg/l (Davitaya, 1969).

However, not all investigations have demonstrated increases in atmospheric aerosol loadings on the regional or global scale. For example, studies by Hoyt and Fröhlich (1983) at Davos in Switzerland have provided no evidence for long-term changes in atmospheric transmission caused by this mechanism. They believe that trends in aercsol loadings from anthropogenic sources are local rather than regional. They suggest that changes in atmospheric transmission occur in urban and industrial areas, but believe that such changes are too localized to be the cause of hemispheric and global trends in temperature.

If an increase in atmospheric aerosols on the macro scale were to be proven, then they could be said to influence temperatures through their impact on the scattering and absorption of solar radiation. The exact effects, however, are still not clear, for whether added aerosols cause heating or cooling of the earth and atmosphere systems is a function not only of their intrinsic absorption and backscatter characteristics, but also of their location in the atmosphere with respect to such variables as cloud-cover, cloud reflectivity and underlying surface reflectivity (Weare *et al.*, 1974). So, for example, over ice caps 'grey' aerosol particles would warm the atmosphere because they would be less reflective than the white snow surfaces beneath, while over a dark reflecting surface they would reflect a greater amount of radiation, leading to cooling (Peck, 1975). Thus precise quantitative assessment of the effects of increased aerosol content in the atmosphere is hàzardous, but Rasool and Schneider (1971) have suggested that a rise by a factor of 4 or 5 in global aerosol concentrations could be sufficient to lower temperatures by as much as 4.5°C.

Similarly, Idso and Brazel (1978) and Brazel and Idso (1979) point to the two contrasting tendencies of dust: the backscattering effect producing cooling, and the thermal-blanketing effect causing warming. The second of these absorbs some of the earth's thermal radiation that would otherwise escape to space, and then re-radiates a portion of this back to the land surface, raising surface temperatures. They believe that natural dust from volcanic emissions tends to enter the stratosphere (where backscattering and cooling are the prime consequences), while anthropogenic dust more frequently occurs in the lower levels of the atmosphere causing thermal blanketing and warming.

Industrialization is not, however, the sole source of particles in the atmosphere, nor is a change in temperature the only possible consequence. Bryson and Barrels (1967), for example, argue that intensive agricultural exploitation of desert margins, such as in Rajasthan, India, would create a dust pall in the atmosphere by exposing larger areas of surface materials to deflation in dust storms. This dust pall, they believe, would so change atmospheric temperature that convection, and thus rainfall, would be reduced. Observations on dust levels over the Atlantic during the drought years of the late 1960s and early 1970s in the Sahel suggest that the degraded surfaces of that time led to a great (threefold) increase in atmospheric dust (Prospero and Nees, 1977). There is thus the possibility that human-induced desertization generates dust which could in turn increase the degree of desertization by its effect on rainfall levels.

Atmospheric aerosols can be an important source of cloud-condensation nuclei. Over the world's oceans a major source of such aerosols is dimethylsulphide (DMS). They are produced by planktonic algae in sea water and then oxidize in the atmosphere to form sulphate aerosols. Because the albedo of clouds (and thus the Earth's radiation budget) is sensitive to cloud-condensation nuclei density, any factor that controls planktonic algae may have an important impact on climate. The production of such plankton could be affected by water pollution in coastal seas or by global warming (Charlson *et al.*, 1987). Charlson *et al.* (1992) believe that anthropogenically derived sulphate aerosols, through their direct scattering of short-wavelength solar radiation and their modification of the shortwave reflective properties of clouds, could significantly increase planetary albedo, thereby exerting a cooling influence on the planet. They maintain that the perturbation in global climate through this mechanism could be comparable in magnitude to current anthropogenic greenhouse gas forcing, but, of course, opposite in sign.

Any assessment of the likely consequences of human production of aerosols must bear in mind that natural production of aerosol materials is of a very much higher order. Nonetheless the data suggest that on a global scale human activity is now producing 280 million tonnes per year of small particulate matter into the atmosphere, compared to a natural rate of 1250 million tonnes per year, or a ratio of about 4.5:1.

The most catastrophic effects of anthropogenic aerosols in the atmosphere could be those resulting from a nuclear exchange between the great powers. Explosion, fire and wind might generate a great pall of smoke and dust in the atmosphere which would make the world dark and cold. It has been estimated that if the exchange reached a level of several thousand megatons, a 'nuclear winter' would occur in which temperatures over much of the world would be as low as –15° to –25°C (Turco *et al.*, 1983), though recent simulations by Schneider and Thompson (1988) suggest that some previous estimates may have been exaggerated. They suggest that in the northern hemisphere maximum average land surface summertime temperature depressions might be of the order of 5–15°C.

Fears were also expressed that as a result of the severe smoke palls generated by the Gulf War in 1991 there might be severe climatic impacts. The situation is still not clear, but preliminary studies have suggested that because most of the smoke generated by the oil well fires will stay in the lower troposphere and only have a short residence time in the air, the effects will be local (some cooling) rather than global, and that the operation of the monsoon should not be affected to any significant degree (Browning *et al.*, 1991; Bakan *et al.*, 1991). Furthermore, in the event the emissions of smoke particles were less than some forecasters had predicted, and they were also rather less black (Hobbs and Radke, 1992).

Thermal pollution

One consequence of the burning of fossil fuels is the production of heat. This is probably most important on the local scale where it can be identified as the 'urban heat island'. On a broader scale, the amount of energy used by humans has been negligible compared both to the resources of solar energy and to the energy of photosynthesis of plants. In global terms the total amount of heat released by all human activity is roughly 0.01 per cent of the solar energy absorbed at the surface (Kellogg, 1978: 215). Such a small fraction would have a negligible effect on the overall heat balance of the earth.

Vegetation and albedo

Incoming radiation of all wavelengths is partly absorbed and partly reflected. Albedo is the term used to describe the proportion of energy reflected and hence is a measure of the ability of the surface to reflect radiation.

Land-use changes create differences in albedo which have important effects on the energy balance, and hence on the water balance, of an area. Tall rain forest may have an albedo as low as 9 per cent, while the albedo of a desert may be as high as 37 per cent (Table 2.3).

There has been growing interest recently in the possible consequences of deforestation on climate through the effect of

Table 2.3 Albedo values for different land-use types

Surface type	*Location*	*Albedo* (%)
Tall rain forest	Kenya	9
Lake	Israel	11.3
Peat and moss	England and Wales	12
Pine forest	Israel	12.3
Heather morrland	England and Wales	15
Evergreen scrub (maquis)	Israel	15.9
Bamboo forest	Kenya	16
Conifer plantation	England and Wales	16
Citrus orchard	Israel	16.8
Towns	England and Wales	17
Open oak forest	Israel	17.6
Deciduous woodland	England and Wales	18
Tea bushes	Kenya	20
Rough grass hillside	Israel	20.3
Agricultural grassland	England and Wales	24
Desert	Israel	37.3

Source: from miscellaneous data in Pereira, 1973, collated by author.

albedo change. Ground deprived of a vegetation cover as a result of deforestation and overgrazing (as in parts of the Sahel) has a verymuch higher albedo than ground covered in plants. This could affect temperature levels. Satellite imagery of the Sin-ai-Negev region of the Middle East shows an enormous difference in image between the relatively dark Negev and the very bright Sinai-Gaza Strip area. This line coincides with the 1948–49 armistice line between Israel and Egypt and results from different landuse and population pressures. Otterman (1974) has suggested that the albedo affected by land use has produced temperature changes of the order of 5°C.

Charney and others (1975) have argued that the increase in surface albedo, resulting from a decrease in plant cover, would lead to a decrease in the net incoming radiation, and an increase in the radiative cooling of the air. Consequently, they argue, the air would sink to maintain thermal equilibrium by adiabatic compression. and cumulus convection and its associated rainfall would be suppressed. A positive feedback mechanism would appear at this stage, for the lower rainfall would in turn adversely affect plants and lead to a further decrease in plant cover. However, this view is disputed by Ripley (1976) who suggests that Charney and his colleagues, while considering the impact of vegetative changes on albedo, have completely ignored the effect of vegetation on evapotranspiration. He points out that vegetated surfaces are usually cooler than bare ground since much of the absorbed solar energy is used to evaporate water, and conclude from this that protection from overgrazing and deforestation might, in contrast to Charney's views, be expected to lower surface temperatures and thereby reduce, rather than increase, convection and precipitation.

Removal of humid tropical rain forests has also been seen as a possible mechanism of anthropogenic climatic change through its effect on albedo. Potter *et al.* (1975) have proposed the following model for such change:

Deforestation

↓

Increased surface albedo

↓

Reduced surface absorption of solar energy

↓

Surface cooling

↓

Reduced evaporation and sensible heat flux from the surface

↓

Reduced convective activity and rainfall

↓

Reduced release of latent heat, weakened Hadley circulation and cooling in the mid and upper troposphere

↓

Increased tropical lapse rates

↓

Increased precipitation in the latitude bands 5–25°N and 5–25°S, and a decrease in the equator-pole temperature gradient

↓

Reduced meridional transport of heat and moisture out of equatorial regions

↓

Global cooling and a decrease in precipitation between 45–85°N and 40–60°S

However, more recent studies (Potter *et al.*, 1981) suggest that globally over the past few thousand years the climatic effects of albedo changes wrought by humans have been small and probably undetectable. Similarly, Henderson-Sellers and Gornitz (1984) have sought to model the possible future effects of albedo changes produced by humans and also predict that there will be but little alteration brought about by current levels of tropical deforestation.

On the other hand, Lean and Warrilow (1989) used a general circulation model (GCM) which showed greater changes than previous models and suggested that Amazon basin deforestation would, through the effects of changes in surface roughness and albedo, lead to reductions in both precipitation and evaporation. Likewise a UK Meteorological Office GCM indicates that the deforestation of both Amazonia and Zaire would by changing surface albedo cause a decrease in precipitation levels (Mylne and Rowntree, 1992).

Budyko (1974) believes that the present extension of irrigation to about 0.4 per cent of the earth's surface (1.3 per cent of the land surface) would decrease the albedo, possibly on average by 10 per cent. The corresponding change of the albedo of the entire earth-atmosphere system would amount to about 0.03 per cent; enough, according to Budyko, to maintain the global mean temperature at a level nearly 0.1°C higher than it would otherwise be.

Forests, irrigation and rain

The belief that forests can increase precipitation levels has a long history (Thornthwaite, 1956), and it has been the basis of action programmes in many lands. For example, the American Timber Culture Act of 1873 was passed in the belief that if settlers were induced to plant trees on the Great Plains and prairies, precipitation would be increased sufficiently to eliminate the climatic hazards to agriculture. On the other hand, at much the same time, the view was expressed that 'rain follows the plough.' Aughey, working in Nebraska, for example, believed that, after the soil is 'broken,' rain as it falls is absorbed by the soil 'like a huge sponge,' and that the soil gives this absorbed moisture slowly back to the atmosphere by evaporation (cited by Thornthwaite, 1956: 569), and so increases the rainfall.

These two early and contradictory views illustrate the confusion that still surrounds this question today. Forests undoubtedly influence rates of evapotranspiration, the flow of streams, the level of groundwater and microclimates, but there is little reliable evidence to suggest that regional rainfall is either significantly increased by afforestation or decreased by deforestation. Certainly schemes to augment rainfall levels on desert margins by widespread planting of forest belts are likely to achieve relatively little, for the aridity of deserts and their margins is controlled dominantly by the gross features of the general circulation, especially the subsiding air associated with the big high-pressure cells of the subtropics.

Although forests may not necessarily have a proven effect on regional or continental rainfall levels, they are far more effective than other vegetation types at trapping other kinds of precipitation, especially cloud, fog and mist. Hence deforestation or afforestation can affect water budgets through the degree to which they intercept non-rainfall precipitation. For example, in Hawaii, Norfolk Island Pines *(Araucaria heterophylla)* in an area where the annual rainfall is 2600 mm, condensed an additional 760 mm from heavy cloud. In the San Francisco area, Parsons (1960) recorded about 250 mm of drip from a pine tree on the Berkeley Hills during each of four rainless summers. In Japan this phenomenon has been used to good effect: trees are planted along the coast to intercept the inland drift of sea fogs.

There is one other land-use change that may result in measurable changes in precipitation; namely large-scale crop irrigation in semi-arid regions. The High Plains of the USA are normally covered with sparse grasses and have dry soils throughout the summer; evapotranspiration is then very low. In the last four decades irrigation has been developed throughout large parts of the area, greatly increasing summer evapotranspiration levels. Barnston and Schickedanz (1984) have produced strong statistical evidence of warm-season rainfall enhancement through irrigation in two parts of this area: one extending through Kansas, Nebraska and Colorado, and a second in the Texas Panhandle. The largest absolute increase was in the latter area and, significantly, occurred in June, the wettest of the three heavily irrigated months. The effect appears to be especially important when stationary weather fronts occur, for this is a situation which allows for maximum interaction between the damp irrigated surface and the atmosphere. Hail storms and tornados are also significantly more prevalent than over non-irrigated regions (Nicholson, 1988).

The possible effects of water diversion schemes

The levels of the Aral and Caspian Seas in Central Asia are falling, as are watertables all over the wide continental region. There are proposals to divert some major rivers to help overcome these problems. However, this raises difficult questions 'because it appears to touch a peculiarly sensitive spot in the existing climatic regime of the northern hemisphere' (Lamb, 1977: 671). The low-salinity water which forms a 100–200 m upper layer to the Arctic Ocean is in part caused by the input of fresh water from the large Russian and Siberian rivers. This low-salinity water is the medium in which the pack ice at present covering the polar ocean is formed. The tapping of any large proportion of this river flow might augment the area of salt water in the Arctic Ocean and thereby reduce the area of pack ice correspondingly. Temperatures over large areas might rise, which in turn might change the position and alignment of the main thermal gradients in the northern hemisphere and, with them, the jetstream and the development and steering of cyclonic activity. However, assessment of this particular climatic impact is still very largely speculative, and some recent numerical models indicate that the climate of the Arctic will not be drastically affected by river diversions (Semtner, 1984).

Lakes

It has often been implied that the presence of a large body of inland water must modify the climate around its shores, and therefore that artificial lakes have a significant effect on local or regional climates. Climatic changes produced by the construction of a reservoir are the result of a variety of factors (Vendrov, 1965): the creation of a body of water with a large heat capacity that reduces the continentality of the climate; the substitution of a water surface for a land surface and the rise of the groundwater level in the littoral zone supplying moisture to the evaporating surface (leading to a rise in wind velocity above the lake and in the littoral zone). Schemes have been put forward for augmenting desert rainfall by flooding desert basins in the Sahara, Kalahari and Middle East (see, for example, Schwarz, 1923). However, whether evaporation from lake surfaces can raise local precipitation levels is open to question, for precipitation depends more on atmospheric instability than upon the humidity content of the air. Moreover, most lakes are too small to affect the atmosphere materially in depth, so that their influence falls heavily under the sway of the regional circulation. In addition, one needs to remember that some of the world's driest deserts occur along coastlines. Thus a relatively small artificial lake would be even more impotent in creating rainfall (see Crowe, 1971: 443–50).

However, the climatic effect of artificial lakes is evident in other ways, notably in terms of a local reduction in frost hazard. In the case of the Rybinsk reservoir (*c*.4500 km^2) in the CIS, it has been calculated that the climatic influence extends 10 km from the lake and that the frost-free season has been extended by 5–15 days on average (D'Yakanov and Reteyum, 1965).

Urban climates

It has been said that 'the city is the quintessence of man's capacity to inaugurate and control changes in his habitat' (Detwyler and Marcus, 1972). One way in which such control becomes evident is in a study of urban climates (Landsberg, 1981). Individual urban areas can at times, with respect to their weather, 'have similar impacts as a volcano, a desert, and as an irregular forest' (Changnon, 1973: 146). Some of the changes that can result are listed in Table 2.4.

Table 2.4 (a) **Average changes in climatic elements caused by cities**

Element	*Parameter*	*Urban compared with rural (–, less; +, more)*
Radiation	On horizontal surface	–15%
	Ultraviolet	–30% (winter); –5% (summer)
Temperature	Annual mean	+0.7°C
	Winter maximum	+1.5°C
	Length of freeze-free season	+2 to 3 weeks (possible)
Wind speed	Annual mean	–20 to –30%
	Extreme gusts	–10 to –20%
	Frequency of calms	+5 to 20%
Relative humidity	Annual mean	–6%
	Seasonal mean	–2% (winter); –8% (summer)
Cloudiness	Cloud frequency + amount	+5 to 10%
	Fogs	+100% (winter); –30% (summer)
Precipitation	Amounts	+5% to 10%
	Days	+10%
	Snow days	–14%

Table 2.4 (b) **Effect of city surfaces**

Penomenon	*Consequence*
Heat production (the heat island)	Rainfall + Temperature +
Retention of reflected radiation by high walls and dark-coloured roofs	Temperature +
Surface roughness increase	Wind– Eddying +
Dust increase (the dust dome)	Fog + Rainfall + (?)

Source: H. Landsberg in Griffiths, 1976, p. 108.

Compared with rural surfaces, city surfaces (Table 2.4b) absorb significantly more solar radiation, because a higher proportion of the reflected radiation is retained by the high walls and dark-coloured roofs of the city streets. The concreted city surfaces have both great thermal capacity and conductivity, so that heat is stored during the day and released by night. By contrast the plant cover of the countryside acts like an insulating blanket, so that rural areas tend to experience relatively lower temperatures by day and night, an effect enhanced by the evaporation and transpiration taking place.

Another thermal change in cities, contributing to the development of the 'urban heat island,' is the large amount of artificial heat produced by industrial, commercial and domestic users.

In general the highest temperature anomalies are associated with the densely built-up area near the city centre, and decrease markedly at the city perimeter. Observations in Hamilton, Ontario, and Montreal, Quebec, suggested temperature changes of 3.8 and 4.0°C respectively per kilometer (Oke, 1978). Temperature differences also tend to be highest during the night. The form of the urban temperature effect has often been likened to an 'island' protruding distinctly out of the cool 'sea' of the surrounding landscape. The rural-urban boundary exhibits a steep temperature gradient or 'cliff' to the urban heat island. Much of the rest of the urban area appears as a 'plateau' of warm air with a steady but weaker horizontal gradient of increasing temperature towards the city centre. The urban core may be a 'peak' where the urban maximum temperature is found. The difference between this value and the background rural temperature defines the *urban heat island intensity* ΔT_{u-r} (max) (Oke, 1978: 225).

Table 2.5 Annual mean urban-rural temperature differences of cities

City	*Temperature differences* (°C)
Chicago, USA	0.6
Washington DC, USA	0.6
Los Angeles, USA	0.7
Paris, France	0.7
Moscow, Russia	0.7
Phildelphia, USA	0.8
Berlin, Germany	1.0
New York, USA	1.1
London, UK	1.3

Source: from data in J. T. Peterson and other sources in Detwyler, 1971.

Table 2.5 lists the average annual urban-rural temperature differences for several large cities. Values range from 0.6 to 1.3°C. The relationship between city size and urban-rural difference, however, is not necessarily linear; sizeable nocturnal temperature contrasts have been measured even in relatively small cities. Factors such as building density are at least as important as city size, and high wind velocities will tend to eliminate the heat island effect.

Nonetheless, Oke (1978: 257) has found that there is some relation between heat-island intensity and city size. Using population as a surrogate of city size ΔT_{u-r} (max) is found to be proportional to the log of the population. Other interesting results of this study include the tendency for quite small centres to have a heat island, the observation that the maximum thermal modification is about 12°C, and the recognition of a difference in slope between the North American and the European relationships. The explanation for this last result is not clear, but it may be related to the fact that population is a surrogate index of the central building density.

In many older towns and cities in Western Europe and North America, the process of 'counter-urbanization' has in recent years led to a decline in population, and it is worth considering whether this is reflected in a decline in the intensity of urban heat islands. One attempt to do this, in the context of London (Lee, 1992), revealed the perplexing finding that the heat-island intensity has decreased by day, but increased by night. The explanation that has been tentatively advanced to explain this is that there has been a decrease in the receipt of daytime solar radiation as a result of vehicular atmosphere pollution, whereas at night the presence of such pollution absorbs and re-emits significant amounts of outgoing terrestrial radiation, maintaining higher urban nocturnal minimum temperatures.

The existence of the urban heat island has a number of implications; city plants bud and bloom earlier, some birds are attracted to the thermally more favourable urban habitat, humans find the added warmth stressful if the city is already situated in a warm area, less winter space-heating is required, but, conversely, more summer air-conditioning is necessary.

The urban-industrial effects on clouds, rain, snowfall and associated weather hazards such as hail and thunder are harder to measure and explain than the temperature changes (Darungo *et al.*, 1978). The changes can be related to various influences (Changnon, 1973: 143):

thermally induced upward movement of air
increased vertical motions from mechanically-induced turbulence
increased cloud and raindrop nuclei
industrial increases in water vapour.

Table 2.6 illustrates the differences in summer rainfall, thunderstorms and hailstorms between various rural and urban areas in the USA. These data indicate that in cities rainfall increases ranged from 9 to 27 per cent, the incidence of the thunderstorms increased by 10 to 42 per cent, and hailstorms increased by 67 to 430 per cent.

An interesting example of the effects of major conurbations on precipitation levels is provided by the London area. In this case it seems that the mechanical effect of the city was dominant in

Table 2.6 Areas of maximum increases (urban–rural difference) in summer rainfall and severe weather events for eight American cities

	Rainfall		*Thunderstorms*		*Hailstorms*	
City	*%*	*Location**	*%*	*Location**	*%*	*Location**
St. Louis	+15	B	+25	B	+276	C
Chicago	+17	C	+38	A,B,C	+246	C
Cleveland	+27	C	+42	A,B	+ 90	C
Indianapolis	0	–	0	–	0	–
Washington DC	+ 9	C	+36	A	+ 67	B
Houston	+ 9	A	+10	A,B	+430	B
New Orleans	+10	A	+27	A	+350	A,B
Tulsa	0	–	0	–	0	–

* A = within city perimeter
B = 8–24 km downwind
C = 24–64 km downwind

Source: after Changnon, 1973, p. 144.

creating localized maxima of precipitation both by being a mechanical obstacle to air flow, on the one hand, and by causing frictional convergence of flow, on the other (Atkinson, 1975). A long-term analysis of thunderstorm records for South-East England is highly suggestive – indicating the higher frequencies of thunderstorms over the conurbation compared to elsewhere (Atkinson, 1968). The similarity in the morphology of the thunderstorm isopleth and the urban area is striking. Moreover, Brimblecombe (1977) shows a steadily increasing thunderstorm frequency as the city has grown.

Similarly the detailed Metromex investigation of St. Louis in the USA (Changnon, 1978) shows that in the summer the city affects precipitation and other variables within a distance of 40 km. Increases were found in various thunderstorm characteristics (about +10 to +115 per cent), hailstorm condition (+3 to +330 per cent), various heavy rainfall characteristics (+35 to +100 per cent) and strong gusts (+90 to +100 per cent).

Two main factors are involved in the effect that cities have on winds: the rougher surface they present in comparison with rural areas; and the frequently higher temperatures of the city fabric.

Buildings, especially those in cities with a highly differentiated skyline, exert a powerful frictional drag on air moving over and around them (Chandler, 1976). This creates turbulence, with characteristically rapid spatial and temporal changes in both direction and speed. The average speed of the winds is lower in built-up areas than over rural areas, but Chandler found that in London, when winds are light, speeds are greater in the inner city than outside, whereas the reverse relationship exists when winds are strong. The overall annual reduction of wind speed in Central London is about 6 per cent, but for the higher velocity winds (more than 1.5 m s^{-1}) the reduction is more than doubled.

Studies in both Leicester and London, England, have shown that on calm, clear nights, when the urban heat-island effect is at its maximum, there is a surface inflow of cool air towards the zones of highest temperatures. These so-called 'country breezes' have low velocities and become quickly decelerated by intense surface friction in the suburban areas. A practical implication of these breezes is that they transport pollution from the outer parts of an urban area into the city centre, accentuating the pollution problem during smogs.

Smoke haze and photochemical smog

The emission of miscellaneous pollutants, especially into the atmospheres of industrial cities, increases the number of days with smoke-haze conditions as in eighteenth- and nineteenth-century London. This is also illustrated by the data for some Illinois cities. Chicago, the only major city in the group, was the first to show a sizeable increase, though the moderately industrialized cities of Moline, Springfield and Peoria also show considerable increases beginning in the 1930s. Cairo, a non-industrial city with minimal twentieth-century growth, has had no great temporal increase in smoke-haze day frequencies (Changnon, 1973). In general, Landsberg (1970) estimates that on average the number of particles present over urban areas is ten times greater than that over rural environs. The degree of pollution increases with city size (Table 2.7). Many of these atmospheric particulates are hygroscopic and so tend to promote fog formation as water vapour readily condenses on them. However, although fog generally occurs, more frequently in metropolitan areas, this is not true for very dense fog, for the heat-island effect of greater warmth in the inner city often prevents the thickest night fogs from reaching the densities reported in the suburbs and outlying districts (see data for London, in Table 2.8).

Table 2.7 Air pollution mechanisms

(a) **Urban-rural air pollution gradients** (*concentrations in* $\mu g/m^3$)

Pollution type	*Urban*	*Non-urban* *Proximate*	*Intermediate*	*Remote*
Suspended participates	102.0	45.0	40.0	21.0
Benzene soluble organics	6.7	2.5	2.2	1.1
Ammonium	0.9	1.22	0.28	0.15
Nitrate	2.4	1.40	0.85	0.46
Sulphate	10.1	10.0	5.29	2.51
Copper	0.16	0.16	0.08	0.06
iron	1.41	0.56	0.27	0.15
Manganese	0.07	0.02	0.01	0.01
Nickel	0.02	0.01	0.00	0.00
Lead	1.11	0.21	0.10	0.00

(b) Air pollution related to city size

Population class	*Concentration* ($\mu g/m^3$) *Total suspended particulates*	SO_2	NO_2
Non-urban	25	10	33
< 10,000	57	35	116
10,000	81	18	64
25,000	87	14	63
50,000	118	29	127
1,00,000	95	26	114
4,00,000	100	28	127
7,00,000	101	29	146
10,00,000	134	69	163
30,00,000	120	85	153

Source: Berry, 1974, modified by author.

Table 2.8 Fog frequencies in London

Location	*Hours per year (based on four observations per day) with visibility less than*			
	40 m	*200 m*	*400 m*	*1000 m*
Kingsway (Central London)	19	126	230	940
Kew (inner suburbs)	79	213	365	633
London Airport (outer suburbs)	46	209	304	562
South-East England (mean of 7 stations outside London)	20	177	261	494

Source: after Chandler, 1965.

Local dust plumes can create changes in precipitation. A notable example of this comes from a coal-fired power station near Boulder, Colorado where fly-ash aerosols inadvertently seeded supercooled fog and induced local snowfall. The shape of the plume and the area of snowfall were the same (Darungo *et al.*, 1978).

Measures can be taken to control smoke haze, and Brimblecombe (1987) provides a detailed and scholarly history of atmospheric pollution and control in 'the big smoke' — London. Edward I ordered a man to be tortured for burning coal and fouling the air in the fourteenth century, while more recently the Clean Air Acts in Britain in 1956 and 1963 have led to a reduction in smog in major cities. In London, emissions of smoke had, by 1970, been reduced to one-tenth of what they were in 1956, and in Manchester fog and SO_2 levels decreased, while sunshine levels went up (see Wood *et al.*, 1974).

Because of the presence of particulate matter, the duration of bright sunshine shows significant reductions over urban areas, and in large cities a progressive decline can be traced towards the city centre. Over the period 1921–50, for example, the mean daily sunshine duration outside London was 4.33 hours, whereas in the central area it was 3.60 hours (Chandler, 1965). Since that time, on account of clean-air legislation, the situation has improved. Improvements have also been made with respect to smog pollution in Los Angeles and other great cities. New legislation to reduce the adverse effects of exhausts from cars caused the average monthly oxidant concentration in central Los Angeles to fall from 0.27 ppm in 1965 to 0.17 ppm in 1974. Annual maximum hourly ozone concentrations have also fallen markedly and a Stage 3 Emergency Pollution Alert has not been called in Los Angeles since 1973. Even so, pollution remains a problem of some gravity in the conurbation and it is commonplace for the 0.10 ppm threshold for adverse health effects to be exceeded on most days of the month between May and September (Elsom, 1987: 35).

The downward trend in fog frequency since the introduction of pollution control measures is also demonstrated clearly for Oxford, Central England (Gomez and Smith, 1984), where since the early 1960s fog has only been about half as frequent as it was in the preceding decades.

Smog produced by the burning of coal is often sulphurous. The sulphur dioxide (SO_2) generated can oxidize in air to produce sulphur trioxide (SO_3), which reacts with water vapour (H_2O) in the presence of catalysts to form sulphuric-acid mist (H_2SO4). The precipitation of such acid droplets with sulphate particles creates the so-called 'acid rain' which is liable to contaminate rivers and lakes. The acidification of fresh waters in Scandinavia may be caused by emissions from the industrial heartland of Western Europe, including Britain.

While smog produced by burning coal is common, there is a second widespread type: photochemical smog. The name originates from the fact that most of the less desirable properties of such fog result from the products of chemical reactions induced by sunlight (Cadle, in Matthews *et al.*, 1971: 340ff.).

Unburnt hydrocarbons play a major role in smog formation and result from evaporation of solvents and fuels, as well as incomplete combustion of fossil fuels. In the presence of oxides of nitrogen, strong sunlight and stable meteorological conditions, complex chemical reactions occur, forming a family of peroxyacyl nitrates (sometimes collectively abbreviated to PANs).

Photochemical smog appears 'cleaner' than other kinds of fog in the sense that it does not contain the very large particles of soot that are so characteristic of smog derived from coal-burning.

However, the eye irritation and damage to plant leaves it causes make it unpleasant. Photochemical smog occurs particularly where there is large-scale combustion of petroleum products, as in car-dominated cities like Los Angeles. Its unpleasant properties include a high lead content; also a series of chemical reactions are triggered by sunlight. For example, a photochemical decomposition of nitrogen dioxide into nitric oxide and atomic oxygen occurs, and the atomic oxygen can react with molecular oxygen to form ozone. Further ozone may be produced by the reaction of atomic oxygen with various hydrocarbons.

Photochemical smogs are not universal. Because sunlight is a crucial factor in their development they are most common in the tropics or during seasons of strong sunshine. Their especial notoriety in Los Angeles is due to a meteorological setting dominated at times by subtropical anti-cyclones with weak winds, clear skies and a subsidence inversion, combined with the general topographic situation and the high vehicle density (>1500 vehicles/km^2). However, photochemical ozone pollution can, on certain summer days, reach appreciable levels even in the UK (Colbeck, 1988), especially in large cities such as London (Table 2.9).

Table 2.9 Annual maximum hourly mean ozone concentrations at County Hall, Westminster in Greater London, 1975–85

Year	*Peak value (ppm)*	*Date of peak*	*No. of days ³ 0.080 ppm*
1975	0.150	26 June	16
1976	0.212	27 June	26
1977	0.087	3 July	1
1978	0.103	29 July	2
1979	0.153	27 July	5
1980	0.116	27 August	2
1981	0.112	5 September	6
1982	0.091	3 August	1
1983	0.099	16 July	11
1984	0.088	6 July	3
1985	0.098	25 July	3

Source: Ball and Laxen, 1986.

Rainmaking and other methods of deliberate climatic modification

The most fruitful human attempts to augment natural precipitation have been through cloud seeding. Rainmaking experiments of this type are based on three main assumptions (Chorley and More, 1967: 159).

1. Either the presence of ice crystals in a supercooled cloud is necessary to release snow and rain or the presence of comparatively large water droplets is necessary to initiate the coalescence process.
2. Some clouds precipitate inefficiently or not at all, because these components are naturally deficient.
3. The deficiency can be remedied by seeding the clouds artificially, either with solid carbon dioxide (dry ice) or silver iodide, to produce crystals, or by introducing water droplets or large hygroscopic nuclei.

These methods of seeding are not universally productive. Under conditions of orographic lift and in thunderstorm cells, when nuclei are insufficient to generate rain by natural means, some augmentation may be attained, especially if cloud temperatures are of the order of –10 to –15°C. The increase of precipitation gained under favourable conditions may be of the order of 10–20 per cent in any one storm. In lower latitudes, where cloud-top temperatures frequently remain above 0°C, silver-iodide or dry-ice seeding is not applicable. Therefore alternative methods have been introduced whereby small water droplets of 50 mm diameter are sprayed into the lower layers of deep clouds, so that the growth of cloud particles will be stimulated by coalescence.

Other techniques of warm cloud seeding include the feeding of hygroscopic particles into the lower air layers near the updraught of a growing cumulus cloud (Breuer, 1980).

The results of the many experiments now carried out on cloud seeding are still controversial, very largely because we have an imperfect understanding of the physical processes involved. This means that the evidence has to be evaluated on a statistical rather than a scientific basis so, although precipitation may occur after many seeding trials, it is difficult to decide to what extent artificial stimulation and augmentation is responsible. It also needs to be remembered that this form of planned weather modification applies to small areas for short periods. As yet, no means exist to change precipitation appreciably over large areas on a sustained basis.

Other types of deliberate climatic modification have also been attempted (Hess, 1974). For example, it has been thought that the production of many more hailstone embryos by silver-iodide seeding will yield smaller hailstones which would both be less damaging and more likely to melt before reaching the ground. Some results in Russia have been encouraging, but one cannot exclude the possibility that seeding may sometimes even increase hail damage (Atlas, 1977). Similar experiments have been conducted in lightning suppression. The concept here is to produce in a thundercloud, again by silver-iodide seeding, an abnormal abundance of ice crystals that would act as added corona points and thus relieve the electrical potential gradient by corona discharge before a lightning strike could develop (Panel on Weather and Climate Modification, 1966: 4–8).

Hurricane modification is perhaps the most desirable aim of those seeking to suppress severe storms, because of the extremely favourable benefit-to-cost ratio of the work. The principle once again is that of introducing freezing nuclei into the ring of clouds around the hurricane centre to trigger the release of the latent heat of fusion in the eyewall cloud system which, in turn, diminishes the maximum horizontal temperature gradients in the storm, causing a hydrostatic lowering of the surface temperature. This eventually should lead to a weakening, of the damaging winds (Smith, 1975: 212). A 15 per cent reduction in maximum winds is theoretically possible, but as yet work is largely at an experimental stage. A 30 per cent reduction in maximum winds was claimed following seeding of Hurricane Debbie in 1969 but US attempts to seed hurricanes were discontinued in the 1970s following the development of new computer models of the effects of seeding on hurricanes. The models suggested that although maximum winds might be reduced by 10–15 per cent on average, seeding may increase the winds just outside the region of maximum winds by 10–15 per cent, may increase rainfall in the outer regions of the storms, may either increase or decrease the maximum storm surge, and may or may not affect the direction of the storm (Sorkin, 1982). Seeding of hurricanes is unlikely to recommence until such uncertainties can be resolved.

Much research has been devoted in Russia to the possibility of removing the Arctic Sea ice and to ascertain effects of such action on the climate of northern areas (see Lamb, 1977: 660). One proposal was to dam the Bering Strait, thereby blocking off water flow from the Pacific. The assumption was that more, and warmer, Atlantic water would be drawn into the Central Arctic, improving temperature conditions in that area. Critics have pointed to the possibility of adverse changes in temperatures elsewhere, together with undesirable change in precipitation character and amount.

Fog dispersal, vital for airport operation, is another aim of weather modification. Seeding experiments have shown that fog consisting of supercooled droplets can be cleared by using liquid propane or dry ice. In very cold fogs this seeding method causes rapid transformation of water droplets into ice particles. Warm fogs with temperatures above freezing point occur more frequently than supercooled fogs in mid-latitudes and are more difficult to disperse. Some success has been achieved using sodium chloride and other hydroscopic particles as seeding agents but the most effective method is to evaporate the fog. The French have developed the 'turboclair' system in which jet engines are installed alongside the runway at major airports and the engines produce short bursts of heat to evaporate the fog and improve visibility as an aircraft approaches (Hess, 1974).

In regions of high temperature, dark soils then become overheated, and the resultant high evapotranspiration rates lead to moisture deficiencies. Applications of white powders (India and Israel) or of aluminium foils (Hungary) increase the reflection from the soil surface and reduce the rate at which insulation is absorbed. Temperatures of the soil surface and sub-surface are lowered (by as much as 10°C), and soil moisture is conserved (by as much as 50 per cent).

The planting of windbreaks is an even more important attempt by humans to modify local climate deliberately. Shelterbelts have been in use for centuries in many windswept areas of the globe, both to protect soils from blowing and to protect the plants from the direct effects of high velocity winds. The size and effectiveness of the protection depends on their height, density, shape and frequency. However, the belts may have consequences for microclimate beyond those for which they were planted. Evaporation rates are curtailed; snow is arrested, and its melting waters are available for the fields; but the temperatures may become more extreme in the stagnant space in the lee of the belt, creating an increase in frost danger.

Traditional farmers in many societies have been aware of the virtues of microclimatic management (Wilken, 1972). They manage shade by employing layered cropping systems or by covering the plant and soil with mulches; they may deliberately try to modify albedo conditions. Tibetan farmers, for example, reportedly throw dark rocks on to snow-covered fields to promote late spring melting; and in the Paris area of France some very dense stone walls were constructed to absorb and radiate heat.

Air pollution: some further effects

This chapter has already made much reference to the ways in which humans have changed the turbidity of the atmosphere and the gases within it. However, the consequences of air pollution go further than either their direct impact on human health or their impact on local, regional and global climates.

First of all, the atmosphere acts as a major channel for the transfer of pollutants from one place to another, so that some harmful substances have been transferred long distances from their sources of emission. DDT is one example; lead is another. Thus, since the start of the Industrial Revolution, the lead content of the Greenland ice cap, although far removed from the source of the pollutant (which is largely derived from either industrial or automobile emissions) has risen very substantially. The same applies to its sulphate content. An analysis of pond sediments from a remote part of North America (Yosemite) indicates that lead levels have also been raised as a result of human activities, being more than twenty times the natural levels. The lead, which came in from atmospheric sources, shows a fivefold elevation in the plants of the area and a fiftyfold elevation in the animals compared with natural levels (Shirahata *et al.*, 1980). Some estimates have also compared the total quantities of heavy metals that humans are releasing into the atmosphere with emissions from natural sources (Nriagu, 1979). The increase is eighteenfold for lead, ninefold for cadmium, sevenfold for zinc, and threefold for copper.

There are, it must be stated, signs that as a result of pollution control regulations some of these trends are now being reversed. Boutron *et al.* (1991), for example, have analysed ice and snow that has accumulated over Greenland in the last two decades and find that lead concentrations have decreased by a factor of 7.5 since 1970. They attribute this to a curbing of the use of lead additives in petrol. Over the same period cadmium and zinc concentrations have decreased by a factor of 2.5.

A second example of the possible widespread and ramifying ecological consequences of atmospheric pollution is provided by 'acid rain' (Likens and Bormann, 1974). Acid rain is rain which has a pH of less than 5.65, this being the pH which is produced by carbonic acid in equilibrium with atmospheric CO_2. In many parts of the world, rain may be markedly more acid than this normal, natural background level. Snow and rain in the North-East USA have been known to have pH values as low as 2.1, while in Scotland in one storm the rain was the acidic equivalent of vinegar (pH 2.4). In the Eastern USA the average annual precipitation acidity values tend to be around pH 4, and the degree of acidulation appears to have been increasing.

It needs to be remembered that not all environmental acidification is caused by acid rain in the narrow sense. Acidity can reach the ground surface without the assistance of water droplets. This is as particulate matter and is termed 'dry deposition.' Furthermore, there are various types of 'wet precipitation' by mist, hail, sleet or snow, in addition to rain itself. Thus some people prefer the term 'acid deposition' to 'acid rain.' The acidity of the precipitation in turn leads to greater acidity in rivers and lakes. Studies by Beamish *et al.* (1975) demonstrated that between 1961 and 1975 pH had declined by 0.13 pH units per year in George Lake, Canada, and a comparable picture emerges from Sweden, where Almer *et al.* (1974) found that the pH in some lakes had decreased by as much as 1.8 pH units since the 1930s.

Very long-term records of lake acidification can provide dramatic evidence for a recent magnification of the acid deposition problem. These are obtained by extracting cores from lake floors and analysing their diatom assemblages at different levels. The diatom assemblages reflect water acidity levels at the time they were living. Two studies serve to illustrate the trend. In South-West Sweden (Renberg and Hellberg, 1982) for most of post-Glacial time (that is, the last, 12, 500 years) the pH of the lakes appears to have decreased gradually from around 7.0 to about 6.0 as a result of natural ageing processes. However, over the last few decades, and especially since the 1950s, a further, more marked decrease occurred to present-day values of about 4.5. In Britain, the work of Battarbee and collaborators (1988a, b) shows that at sensitive sites pH values before around 1850 were close to 6.0 and that since then pH declines have varied between 0.5 and 1.5 units.

Fortunately there is some evidence that if industrially derived sulphate emissions decline lake acidity once again becomes reduced. Acidification of lakes thus appears to be a reversible process. This has been demonstrated through a study of lake water chemistry and diatom stratigraphy undertaken in Scotland by Battarbee *et al.* (1988b).

The causes of acid deposition are the increasing quantities of sulphur oxides and nitrogen oxides being emitted from fossil-fuel combustion. At present, roughly 60–70 per cent of the problem is due to sulphur-oxide emissions, and the balance to the nitrogen oxides. However, in some regions, such as the west coast of the United States and Japan, the nitric acid contribution may well be of greater relative importance. Moreover, with the emissions of oxides of nitrogen appearing to increase at a faster rate than sulphur-dioxide emissions, nitric acid will probably increase further in its relative importance.

Two main factors contribute to the increasing seriousness of the problem. One is the replacement of coal by oil and natural gas. The second, paradoxically, is a result of the implementation of air pollution control measures (particularly increasing the height of smoke stacks and installing particle precipitators). These appear to have transformed a local 'soot problem' into a regional 'acid rain problem'. Coal-burning produced a great deal of sulphate, but was largely neutralized by high calcium contents in the relatively unfiltered coal smoke emissions. Natural gas burning creates less sulphate but that which is produced is not neutralized. The new higher chimneys pump the smoke so high that it is dispersed over wide areas, whereas previously it returned to earth nearer the source. Some of the stacks are now enormous. One at the Sudbury copper-nickel smelter in Canada is 400 m tall. A possible subsidiary reason for some of the observed increases in acidity is that there may now be less calcium-rich dust in the atmosphere over North America since the passing of the Dust Bowl years (Stensland and Semonin, 1982).

The effects of acid rain are especially serious in areas underlain by highly siliceous types of bedrock (for example, granite, some gneisses, quartzite and quartz, sandstone), such as the old shield areas of the Fenno-Scandian shield in Scandinavia and the Laurentide shield in Canada (Likens *et al.*, 1979).

The ecological consequences of acid rain are still the subject of some debate. Krug and Frink (1983) have argued that acid rain only accelerates natural processes, and point out that the results of natural soil formation in humid climates include the leaching of nutrients, the release of aluminium ions and the acidification of soil and water. They also note that acidification by acid rain may be superimposed on longer-term acidification induced by changes in land use. Thus the regrowth of coniferous forests in what are now marginal agriculture areas, such as New England and highland Western Europe, can increase acidification of soils and water. Similarly, it is possible, though in general unproven (see Battarbee *et al.*, 1985), that in areas like Western Scotland, a decline in upland agriculture and the regeneration of heathland could play a role in increasing soil and water acidification. Likewise Johnston *et al.* (1982) have suggested that acid rain can cause either a decrease or an increase in forest productivity, depending on local factors. For example, in soils where cation nutrients are abundant and sulphur or nitrogen are deficient, moderate inputs of acid rain are very likely to stimulate forest growth.

In general, however, it is the negative consequences of acid rain that have been stressed. One harmful effect is a change in soil character. The high concentration of hydrogen ions in acid rain causes accelerated leaching of essential nutrients, making them less available for plant use. Furthermore, the solubility of aluminium and heavy metal ions increases, and instead of being fixed in the soil's sorption complex, these toxic substances become available for plants or are transferred into lakes, where they become a major physiological stress for some aquatic organisms.

Freshwater bodies with limited natural cations are poorly buffered and thus vulnerable to acid inputs. The acidification of thousands of lakes and rivers in Southern Norway and Sweden during the past two decades has been attributed to acid rain, and this increased acidity has resulted in the decline of various species of fish, particularly trout and salmon. But fish are not the only aquatic organisms that may be affected. Fungi and moss may proliferate, organic matter may start to decompose less rapidly, and the number of green algae may be reduced.

Forest growth can also be affected by acid rain, though the evidence is not necessarily proven. Acid rain can damage foliage, increase susceptibility to pathogens, affect germination and reduce nutrient availability. However, since acid precipitation is only one of many environmental stresses, its impact may enhance, be enhanced by, or be swamped by other factors. For example, Blank (1985), in considering the fact that an estimated one-half of the total forest area of the former West Germany was showing signs of damage, referred to the possible role of ozone or of a run of hot, dry summers on tree health and growth.

Further useful information on the contentious issue of the effects of acid rain is contained in Likens and Butler (1981), Hutchinson and Havas (1980), Hornbeck (1981), Park (1987) and Wellburn (1988).

A very rapidly developing area of concern in pollution studies is the current status of stratospheric ozone levels. The atmosphere has a layer of relatively high concentration of ozone (O_3) at a height of about 16 and 18 km in the polar latitudes and of about 25 km in equatorial regions. This ozone layer is important because it absorbs incoming solar ultraviolet radiation, thus warming the stratosphere and creating a steep inversion of temperature at heights between about 15 and 50 km. This in turn affects convective processes and atmospheric circulation, thereby influencing global weather and climate. However, the role of ozone in controlling receipts of ultraviolet radiation at the earth's surface has great ecological significance, because it modifies rates of photosynthesis. One class of organism that has been identified as being especially prone to the effects of increased ultraviolet radiation consequent upon ozone depletion are phytoplankton—aquatic plants that spend much of their time near the sea surface and are therefore exposed to such radiation. A reduction in their productivities would have potentially ramifying consequences, because these plants directly and indirectly provide the food for almost all fish.

Human activities appear to be causing ozone depletion in the stratosphere, most notably over the south polar regions, where an 'ozone hole' has been identified. Possible causes of ozone depletion are legion, and include various combustion products emitted from high-flying military and civil supersonic aircraft; nitrous oxide released from nitrogenous chemical fertilizers; and chlorofluorocarbons (CFCs) used in aerosol spray cans, refrigerant systems, and in the manufacture of foam fast-food containers.

However, in recent years the greatest attention has been focused on the role of CFCs, the production of which has been climbing in the last few decades. These gases may diffuse upwards into the stratosphere where solar radiation causes them to become dissociated to yield chlorine atoms which react with and destroy the ozone. The process has been described thus by Titus and Seidel (1986: 4):

> Because CFCs are very stable compounds, they do not break up in the lower atmosphere (known as the troposphere). Instead, they slowly migrate to the stratosphere, where ultraviolet radiation breaks them down, releasing chlorine.
>
> Chlorine acts as a catalyst to destroy ozone; it promotes reactions that destroy ozone without being consumed. A chlorine (Cl) atom reacts with ozone (O_3) to form ClO and O_2. The ClO later reacts with another O_3 to form two molecules of O_2, which releases the Cl atoms. Thus two molecules of ozone are converted to three molecules of ordinary oxygen, and the chlorine is once again free to start the process. A single chlorine atom can destroy thousands of ozone molecules. Eventually, it returns to the troposphere, where it is rained out as hydrochloric acid.

The Antarctic ozone hole has been identified through satellite monitoring and by monitoring of atmospheric chemistry on the ground. The reasons why this zone of ozone depletion is so well developed over Antarctica include the very low temperatures of the polar winter, which seem to play a role in releasing chlorine atoms; the long sunlight hours of the polar summer, which promote photochemical processes; and the existence of a well-defined circulation vortex. This vortex is a region of very cold air surrounded by strong westerly winds, and air within this vortex is isolated from that at lower latitudes, permitting chemical reactions to be contained rather than more widely diffused. No such clearly defined vortex exists in the northern hemisphere, though ozone depletion does seem to have occurred in the Arctic as well (Proffitt *et al.*, 1990). Furthermore, observations in the last few years indicate that the Antarctic ozone hole is spreading over wider areas and persisting longer into the Antarctic summer. It is also possible that the situation could be worsened by omissions of volcanic ash into the atmosphere (as from Mount Pinatubo), for these can also cause chemical reactions that lead to ozone depletion (Mintzer and Miller, 1992).

It is perhaps worth concluding this section by reiterating the fact that the variability of air pollution in both space and time, while undoubtedly owing much to human action, is also highly dependent on natural conditions, and in particular on meteorological conditions (Thompson, 1978). Climatic factors control the efficiency with which effluents are diluted and dispersed away from primary sources, while local relief and microclimatic circumstances modify the interaction between the emission of pollutants and the atmosphere. Moreover, as we have already stated, not all pollutants result from human endeavour: the vast majority of nitrogen oxides are emitted by bacteria rather than by car exhausts, most chlorides in rain water come from the oceans, and a considerable proportion of atmospheric dust is injected into the air by such mechanisms as volcanic eruptions, and by dust storms removing fine silt from glacial outwash fans or desert basins.

□□□

3

Can Ecologism Transform Capitalism?: Sustainable Development, Ecological Modernisation and Economic Democracy

WHAT FORM AND LEVEL of economic activity is permissible if we wish to maintain biodiversity in the present world, thereby satisfying the requirements of ecological justice? This is a world in which there is a huge human population which is rapidly growing. It is one in which the most affluent societies, whose level of *per capita* wealth it appears to be the aspiration of the majority of the world's human inhabitants to attain, have only acquired that wealth by destroying habitats and reducing or exterminating many species. Does this not imply that, in the course of realising that aspiration, the teeming billions of humanity will necessarily destroy more habitats and exterminate many more species, at least up to the point at which the biosphere is so damaged that it is apparent to everyone that no further economic growth is possible without destroying the biological basis of human life itself?

The answer to this question which has been popular among many environmentally concerned people is that this is indeed what is implied if humanity follows its present course. Thus, if we are genuinely concerned to protect biodiversity, for all the reasons, prudential and moral, given earlier, then we have radically to alter our modes of economic activity as well as doing something rapid and effective to halt the escalation of the human population.

In this final part of the book we will be exploring some of the possibilities for the organisation of economic and political life which hold out some promise that human beings can acquire the material and other benefits necessary for their own flourishing in a way that respects the demands of ecological justice. However, this whole discussion is overshadowed by one inescapable fact, namely that the political economy of liberal capitalism, in spite of the vicissitudes with which it is endemically beset, is dominant, and is growing even more dominant, right across the globe.

There are good reasons, which we will shortly be going into, why political ecologists and other environmentally concerned people should regard this fact with serious concern. Their reactions to it divide roughly into two sorts. The pessimists see no real hope of altering capitalism's fundamental nature, and so see no real hope for an environmentally secure future. We are, so to speak, trapped aboard a runaway train which will stop only when it crashes into the environmental buffers.

The more optimistic adopt the approach of the exponent of Judo to the problem, seeking to use capitalism's own tendencies and momentum so to transform it that, in spite of its present tendencies, it will at least begin to conform to the requirements of ecological health and justice. Capitalism may never satisfy these requirements to any great degree, but this approach hopes at least to buy time for an alternative to capitalism to be found.

Political economy: critique and counter-critique

Let us first survey the general debate between environmentally concerned people and their critics concerning matters of political economy in order to set the stage for the rest of the discussion. There are two distinct issues involved in this area. The first is the question of the material viability of our current economic practices, which in turn encompasses two general problems. These are:

(1) Resource problems—Will we have the natural materials to sustain the economic activities of a rapidly growing population? 'Natural materials' encompasses land, water, fuels, ores and so on.

(2) Sink problems—Will we be able to dispose of the waste products of our economic production in a way which does not poison or destroy our sources of clean water, land and air?

The second issue concerns the human significance of our economic activity. For we do not simply produce in order to meet physical needs. In the course of material production, or praxis, we transform our lives and ourselves in significant ways. Human beings exercise their imaginations, their emotions, their moral sense and their religious beliefs in the course of material production. In identifying and overcoming the problems of material production by the invention of new technologies and ways of working, human beings have come to make over the world in a way which renders it more humanly intelligible. On some views, such as those of Hegel, they have, on balance, come to be more 'at home' in the world as a result. If this is a correct analysis, then the question raised by proposals to alter the basis of human economic activity is whether they amount to a viable praxis.

To this list of issues the ecologist adds one more. This is the issue of whether, if we can find a humanly satisfying form of economic activity which avoids or minimizes resource and sink problems and deals with the problem of human poverty, it will also satisfy the requirements of ecological justice.

Problems (1) and (2) were among the first to be recognised and highlighted by environmentalists. An alarming picture emerged from their analyses of human beings as about to run out of material resources and to pollute the planet to a terminal degree. This picture underpinned appeals addressed directly to human self-interest and led to the idea that the system of economic production to which we are apparently committed, far from being conducive to human well-being, would lead in a short time to human misery of untold proportions.

From this diagnosis have emerged well-known environmental themes:

(1) The critique of our present conception of economic growth, based on Gross National Product (GNP) and Gross Domestic Product (GDP), which measures as a

positive contribution to human well-being all the measures we have to take to remedy self-inflicted harms, arising from the profligate and inequitable forms of economic activity in which we engage.

(2) The injunction to cut down on waste, to recycle as much as we can of the materials used in manufacture, and the emphasis on the need to turn to renewable sources of energy and to reduce our consumption of non-renewable fossil fuels and ecologically suspect nuclear power.

(3) The attack on the increase in polluting and greenhouse gases, such as carbon dioxide, released by the use of fossil fuels in the course of our modern manufacturing industries and modes of transportation, coupled with the highlighting of global warming and ozone depletion as examples of harmful large-scale human effects on the biosphere.

(4) The attack on the polluting, ecologically harmful and inhumane methods of modern mass-production agriculture, and the urging of farmers to turn to organic, humane and ecologically beneficial methods for growing our food.

(5) The claim that human numbers and economic activity adversely affect habitats and biodiversity. This adverse impact is said to be substantial purely in terms of what it means for human economic resources, ecological services and the conditions needed for human culture to flourish, even without going into the claims political ecologists also wish to make about the moral considerability of the non-human.

(6) The claim that the enormous increase in human numbers which threatens to enhance the environmentally harmful impact of the contemporary economic system to a disastrous extent is at least partly caused by poverty, which has to be dealt with by a more equal distribution of resources rather than by extension of the pernicious forms of economic activity among the poverty populations of the world.

These claims have all been subject to vigorous counter-critiques by defenders of the existing economic system and economic growth as understood within orthodox economics. Before we look at these criticisms we need first to present, from the environmentalist perspective, the possible causes of the above phenomena.

Controversy has arisen among environmentalists on this matter. Some lay the emphasis on the rise of modern industrialism as the prime culprit in the creation of an ecologically harmful economic system. This ushered in the era of massproduction with its concomitant demand for the extraction and use of enormous amounts of raw materials and fossil fuels. On this view, industrialism, wherever it occurs—under a capitalist, socialist or any other system—is what has to be radically restructured in accordance with ecological imperatives. Fitting in with this view there is the call for the adoption of simpler, less materialistic lifestyles and the turning to more spiritually rewarding activities which involve harmony with, rather than destruction of the environment.

Others, such as David Goldblatt, regard this 'industrialism-plus-materialistic-greed' analysis as being too superficial, and point to the inner dynamic of capitalism, with its ceaseless quest for profit and concomitant urge to create ever new markets and endlessly to stimulate human demands for material goods and services, as the main culprit. To the obvious criticism of this view that the worst environmental records this century are to be found within those societies which abolished capitalism and sought to employ a system of rational planning of economic activity to meet human need directly, Goldblatt replies that such societies were seeking to emulate the practice of capitalism without the democratic safeguards present in capitalist societies. The latter, he argues, have mitigated some of the worst effects of capitalism precisely by permitting the emergence of an environmentally conscious social movement.

To this is added the claim that industrialism on its own would not have had the environmentally harmful effects with which we have become familiar in the late twentieth century. As Goldblatt argues, only when industrialism was organised within the capitalist system which it had itself unleashed did the potential created by the discovery of coal-based forms of energy production to release economic activity from previous ecological limits become a reality.

From this perspective the aim of environmentally concerned people should be to control rigorously the workings of capitalism in the interests of the preservation of the biosphere. This will involve a mixture of state and international regulation, taxation policies, licensing regimes and the increased democratic accountability of capitalist corporations. It may also involve encouraging capitalist firms to engage in the activities which go under the heading of 'ecological modernisation.' This involves, among other things, the manufacture of environmentally beneficial technologies— for example, catalytic converters and other devices to reduce the environmentally harmful effects of economic production and transportation — by processes which are themselves made minimally damaging to the environment.

Having outlined the environmentalist perspectives on matters of political economy, let us now rehearse the counter-criticisms of all that has been said so far offered by those who are sometimes labelled 'contrarians'.

The basic strategy of critics of the environmental views listed above is to claim that where there are genuine problems they can be, and are being, satisfactorily tackled within the economic dispensation as it exists in the late twentieth century, but that most of the 'problems' are in reality non-existent (or, more weakly, that the case that there is a genuine problem has not yet been established). Thus, the resource problems are being dealt with by the usual means of finding alternative raw materials and sources of energy. Human ingenuity is the key to their solution. Solar and other forms of renewable energy, new forms of nuclear energy without the creation of nuclear waste, even the long-sought fusion power, when coupled with already demonstrated ability to use energy more efficiently, will cope with any energy needs we are likely to have, even with a greatly enlarged human population.

New materials and production techniques will cope with any potential resource shortages. Human ingenuity is the crucial resource here once again. Thus, the possibility of nanotechnology, which involves the creation of any material we desire from atoms, promises to cure whatever raw material shortages may threaten, and key steps have already been taken towards it. In any case, much modern technology is based on the computer, the production of which involves a reduction in the amount of material used concomitantly with massive increases in computing power achieved. Further, much economic growth in the future will come from forms of activity which are much less material and energy intensive than the smokestack industries of the early phases of the Industrial Revolution. Information and education are the key to future economic growth, and these use far fewer material resources than their predecessors.

In the area of pollution and waste, critics of environmentalism point to the way in which the pollution-creating activities of units such as factories, powerstations and automobiles have all been drastically reduced in recent years, partly as a result of the recognition that polluting forms of production are usually inefficient forms of production. They emphasise the robustness of the biosphere and its ability to recover very quickly from mishaps (miscalled 'disasters' by environmentalists) such as large oil spills at sea. They point to the way in which alternatives were found to the ozone-destroying CFCs and confidently expect that the ozone layer will rapidly repair itself, or at least not deteriorate further — although this is a problem which they claim was overblown in the first place. They strongly challenge the claim that global warming is being produced as the result of releases of greenhouse gases into the atmosphere in the course of human activities, but argue that, even if it is occurring, there is no reason to suppose that its effects will be sufficiently harmful to warrant strenuous steps to prevent it. In their view we are ignorant of what its effects are likely to be, and for all we do know it is as likely to have beneficial effects overall as it is to be harmful.

In the sphere of agriculture they argue that the problems of pesticide poisoning of the environment have largely been overcome by the invention of transient and more finely tuned pesticides. Genetically engineered plant strains also promise to make the application of pesticides less extensive. They simply reject organic agriculture as a romantic fad, and point to the enormous increases in agricultural productivity which science and fossil-fuel-based agriculture have achieved and which further developments in biotechnology promise to increase in the future. Thus, they argue, there is no reason why the whole of the expected increase in humanity should not be adequately, or even well, fed in the future, in the manner to which the fortunate citizens of the West have become accustomed since the early years of the twentieth century, provided that the advances possible under scientific agriculture married to the market economy are not stymied in the name of a misconceived concern to protect the environment.

With respect to the decline in habitats and the loss of biodiversity, critics are again inclined strongly to be sceptical of the claim that serious losses of species are occurring, or are likely to occur. Even if the claims of environmentalists are correct in this regard, critics of environmental arguments challenge the view that any serious harm to the biosphere will ensue, or that human beings will forego large economic, scientific and medical benefits potentially present in the lost species. They further claim that it is not proven that reduction or fragmentation of habitats result in extinctions—or in large-scale extinctions, at any rate. In any case, humans come first, they argue, and if the cost of creating a decent life for human beings is the loss of some species and habitats, then this is a price well worth paying.

The moral arguments of ecologists are, of course, rejected. Non-humans are held to have only instrumental value, insofar as they are necessary for humanitys economic, aesthetic or, at a pinch, spiritual benefit. The view that non-human creatures are morally considerable and intrinsically valuable is regarded as an affectation of the comfortable Western middle class, reacting to the disenchantment of the world resulting from the success of modern science and their loss of religious faith. It is not to be taken seriously.

Live for a while as a Third World peasant, critics say, facing the perpetual threat of famine, drought, earthquakes, floods, disease and the depredations of wild animals, from bugs to elephants, and the truth about the non-human natural world will quickly become apparent, namely that it is a constant threat to human happiness and well-being. Those human beings may count themselves fortunate whose forebears, by their enormous economic and technological efforts, released their descendants from thraldom to the natural world, allowing them to dwell on its more aesthetically pleasing aspects from the comfort of their totally artifical urban or rural environments based on the subjugation of the natural world to human will. This is the condition to which all human beings, rightly, aspire.

These ripostes to the environmentalists' claims are underpinned by the critics' celebration of economic growth as hitherto understood, and by their emphasis on the vital instrumental importance of capitalism, free trade and the market economy in generating and sustaining that growth. At which point we move on to the more familiar and much-debated issues of the merits of capitalist markets and their relationship to issues of justice, equality and liberty among human beings, an issue which, as we have seen, is of direct interest to environmentalists, but not only to them.

It is clear that this series of disputes between environmentalists and their critics covers issues which are partly empirical and partly concern rival value systems. On the empirical side, the critics are on their strongest ground in their claim that resource problems have been misconceived or exaggerated by environmentalists. Here, human ingenuity is indeed still promising as the key to finding the energy sources and raw materials needed to sustain our material culture.

However, they become less and less convincing as they venture further into fields in which the health of the biosphere is in question. They appear overly sanguine about the possibilities of mass, fossil-fuel-based agriculture, whose monocultural basis remains a hostage to fortune. Their confidence that the problems associated with pesticides have now been overcome are premature, to judge by a 1997 UK report on the harmful indirect effect of temporary, targeted pesticides on British birdlife.

They are correct to note that there have been important steps forward in making units of manufacture and transport less polluting, at least in the West, but what they have not shown is that the benefits of reducing unit pollution will not be swamped by the enormous increase in the sheer number of production and transport units, as the Third World catches up with Western patterns of industrialisation.

They seem to be correct in their view that the depletion of the ozone layer, if it is indeed due to the emission of CFCs, is a problem which is easily curable, at least in principle. But they are beginning to look somewhat cavalier in their nonchalance with respect to the issue of global warming. It is clearly correct that we do not know what its effects, if any, will be, but it is riot unreasonable to suppose that it is occurring, that human activity is involved in its production, that its effects will be very marked, and that it is at least possible that those effects will be far from beneficial. At the very least this looks to be a game of Russian roulette on a massive scale.

They appear to be even more cavalier in their attitude towards biodiversity and habitat destruction. The loss of species referred to by ecologists is not going to result in neat piles of visible corpses for human beings to count easily. The species, especially those located in the all-important rainforest habits, are usually going to be of creatures which are small in size and highly localised. Extinction, even when rapid, need not produce any dramatic visible effects, even for experienced and skilled observers. As ecologists dramatically but correctly point out, the species thereby lost are lost forever. If they disappear before we even find out that they exist then we have no way of even contemplating resurrecting them by the wizardry of biotechnology.

It is correct to point out, as the critics of environmentalism often do, that the evidence shows that, even with dramatic pollution incidents such as massive oil spills, natural processes, when allowed to operate, can bring the affected environment back to something like its previous condition quite quickly. Nature is indeed very robust, as the history of extinction episodes during the last 600 million years has revealed. But, although it is difficult to damage the biosphere beyond repair this does not imply that it is impossible to do so. Although it would be hard for human beings to create as much havoc on the biosphere as was produced by the asteroid whose collision with earth exterminated the dinosaurs and enormous numbers of other life forms, an all-out nuclear war among human beings would not be a bad approximation to it.

What this suggests is that since the 1940s human beings have acquired the possibly unique power to inflict upon the biosphere serious, perhaps terminal, damage. Given our enormous appetite for materials, the huge and rapidly increasing size of our population, unprecedented for a mammal of our dimensions, and our ignorance concerning the precise effects of our activities upon the biosphere, it ill becomes any of us to be other than extremely cautious in our estimate of whether or not the biosphere can take all we can throw at it.

However, the critics clearly have a good point in their emphasis on the fact that for most human beings, for most of their history, the biosphere, besides providing the conditions to sustain humanity in existence, has also constantly threatened it with a whole host of unpleasant phenomena. The protection of ourselves from these threats correctly remains a main priority for human activity. But this has to be balanced against the fact that even Third World peasants show a love of their natural environment, for all its actual and potential harshness, and that in the practical business of conservation there are many people in the Third World who regard it as important, not just for economic reasons. Some such people devote their lives, often at risk to themselves, to its protection.

There exist many Third World environmental protection organisations, often with members drawn from a peasantry which sees its immediate natural environment as under threat from the rapacity and/or thoughtlessness of governments, corporations and powerful individuals. Thus, it is just not true that concern for the welfare of non-humans is merely a Western middle-class fad. To see this one has only to recollect that many religious and moral viewpoints pervasive in the Third World and among aboriginal people in general testify to the importance they attach to a proper respect for our fellow creatures.

One might also add that the threats posed by the natural world are at least matched by the threats which human beings pose to each other in the form of war and genocide, and that often the ill-effects of the non-human world upon human beings is mediated by human malice and stupidity—consider how often famines have occurred in situations where food was in plentiful supply but withheld from those facing starvation.

As far as the attack on the moral views of ecologists is concerned, all that can be done is to pursue the arguments, such as those presented in Parts Two and Three of this book. These may be faulty, but, if they are, that has to be shown to be so by the normal procedures of rational debate, not by the *ad hominem* condemnation of them as the sentimental conceits of middle-class Westerners with nothing better to think about.

Versions of sustainable development

Having argued that the environmentalists are correct to be concerned about the workings of untrammelled capitalism, we need to contemplate some of the ways in which capitalism may be redeemed, environmentally speaking. Three pertinent ideas require discussion here, namely sustainable development, ecological modernisation and economic democracy.

It is by now standard to note that there has been developing since the publication of the Brundtland report in the late 1980s the idea that it may be feasible to attain the economic growth necessary to deal with widespread human poverty in a way which does not involve sawing off the biospherical branch upon which we are sitting. Humanity, it is claimed, has the possibility of instituting a process of 'sustainable development', which involves so acting that we bequeath to our descendants a world which is as good as the world which we have inherited from our forebears.

Whether or not the concept of sustainable development does make sense, it is a tribute to the influence of environmental thinkers that it has been thought necessary in a major United Nations report to try to rethink the basis upon which economic activity is carried out. However, we need to consider whether the idea of sustainable development in some form can commend itself to ecologism, or whether instead political ecologists have to propose a radical abandonment of traditional economic goals. This could encompass the institution of a much lower level of economic activity together with a much simpler and less affluent lifestyle, at least in material terms, and a greater willingness to share what we do possess in the way of material resources.

As Munda has pointed out, we are faced with at least two conceptions of what sustainable development actually is. The view prevalent in the sub-discipline of neo-classical economics called 'environmental economics' is that we should understand sustainable development in accordance with the criterion of 'weak sustainability.' To use Munda's formulation, an economy can be considered weakly sustainable if it saves more than the combined depreciation of natural and man-made capital. As Turner *et al.* put it:

> We can pass on less environment so long as we offset this loss by increasing the stock of roads and machinery, or other man-made (physical) capital. Alternatively, we can have fewer roads and factories as long as we compensate by having more wetlands or mixed woodlands, or more education.

The main objection given by Munda to this conception is that it assumes the perfect substitutability of different forms of capital, natural and man-made. This concept has to be rejected because, as he explains:

(1) Man-made capital is not independent of natural capital: in creating capital, natural resources must be used, so that the attempt to substitute such artificial capital for natural capital will be limited by the extent to which the increase in artificial capital requires an input of natural resources. The only way to avoid this being a problem is to create capital whose productivity outweighs the natural resources used up in its creation, and it cannot just be assumed that this is possible.

(2) Artificial capital is monofunctional (a road exists solely to carry traffic) whereas natural capital is multifunctional (woodland provides food and shelter, acts as a sink for carbon dioxide, is a source of wood).

To this has to be added the important consideration adduced by two economists working within environmental economics, David Pearce and Kerry Turner, who argue that the maintenance of the natural capital stock is a condition for sustainable development. This is because we do not possess full scientific knowledge of how the biosphere works nor of its precise role in supporting human economic systems, thus we must retain it as far as possible, on pain of destroying irreversibly something which may turn out to be essential to our economic and other well-being.

An alternative version of sustainable development is that put forward by the newly developing discipline of ecological economics. This rejects a whole host of the assumptions present in neo-classical environmental economics, such as the possibility of calculating costs and benefits of different economic activities by reducing all factors to a single criterion of comparability, namely the monetary one. Ecological economics argues that sustainable development must employ the criterion of strong sustainability. As Munda characterizes it, this assumes that certain kinds of natural capital are critical to the health of the biosphere and that such capital is not substitutable by means of man-made capital. Instead, we need to develop non-monetary indicators of ecological sustainability, based on the direct physical measurement of important stocks and flows of natural capital.

This will be inherently problematic for many reasons, among which are the fact that such indicators will be subject to scientific uncertainty and to political negotiation, as different actors attempt to persuade decision-makers to adopt their preferred interpretations. Thus, the working out of what is to be done to maintain economic activity across the globe in a strongly sustainable mode requires the development of democratic processes for the articulation and negotiation of incommensurable values and considerations.

If we ask the question which version of sustainable development ecologism should support, it is clear that it has to be that espoused by ecological economics. Defenders of biodiversity on the ground of the intrinsic value of species are *ipso facto* committed to the rejection of the valuing of species in purely monetary terms, that is, they are committed to rejecting the view of species as purely 'natural capital', even though they accept that each species is also potentially a resource for human economic purposes.

Hence, they must be attracted to a view of economic decision-making which allows for the fact that different, mutually incompatible, and irreducible values are legitimately in play. They will also be happiest with a view of sustainable development which emphasises the need to maintain natural capital, for that promises that the case will be more readily acceptable for the preservation of natural habitats within winch biodiversity may be preserved.

Finally, the emphasis which ecological economics places on the centrality of democratic processes of derision-making is one which fits into ecologism's case for democratic decision-making and the guardianship institution outlined in the previous part of this book. The latter wili be a key player in the democratic debates which are envisaged as being central to the decision-making process, even if, there are good reasons for insulating the members of such an institution from some of the direct effects of democratic politics.

However, clearly there are great difficulties involved in working out methods for finding indicators of key elements in the natural environment in order to conduct the debates. It is also clear that ecological economics does not seem to have committed itself to the idea that any group participating within the debate will be accorded a veto power, whereas that is something envisaged by the creation of the guardianship institution. However, it is part of the logic of the non-substitutability of different forms of capital that someone should be able to block proposals to destroy a critical part of the natural capital, especially as such decisions are nearly always non-reversible. Hence, provided that there is scientific knowledge to sustain it, a veto over the destruction of key elements of natural capital is necessary.

A crucial element in the viability of democratic decision-making which aims at sustainability is, as has forcefully been argued by Gray, that the accounting procedures of capitalist corporations must be radically altered to make transparent to democratic decision-makers what precise impact the economic activities of such corporations are having upon the sustainability of the human economy. He argues, with great plausibility, that neither voluntary actions by corporations nor the workings of markets will move far enough and quickly enough in the direction of what he calls 'reporting for sustainability'. At the end of the twentieth century, both markets and corporations respond solely to financial indicators. The fact that their impacts on environmental processes often resist expression in financial terms requires that new forms o reporting of their activities be devised and enforced. Even whei. these are in place, however, capitalist firms will not be in a position themselves to make informed decisions about the sustainability of their activities, nor is it reasonable to require them to take the sole responsibility for acting sustainably. This is a matter for democratic decision-making by the wider society, involving, as it necessarily does, judgements about trade-offs involving a whole variety of factors.

Thus far, the concept of sustainable development, although of considerable interest to ecologism, remains at a rather high level of abstraction. A more specific strategy for 'greening' political economy, specifically in its capitalist form, is that known as ecological modernisation.

Ecological modernisation

The concept of ecological modernisation is simple enough to explain. In Dryzek's formulation it involves the restructuring of the capitalist political economy along more environmentally sound lines. Adverse environmental effects are held to stem from the way that productive and distributional processes are organised within late twentieth-century capitalist economies. No inherent difficulties are believed to stem from the capitalist form itself. The trick is, therefore, to induce the appropriate forms of economic reorganisation within capitalist enterprises in a way which protects profitability while reducing or minimising adverse environmental effects, such as pollution.

Dryzek argues that the capitalist economies which have been most successful so far in this endeavour have been those of Germany, Holland, Japan, Sweden and Norway. They have been conspicuously successful in increasing the energy efficiency of national income (by, for example, changing to more energy-efficient processes), reducing the amount of harmful emissions caused during economic activity and reducing the amount of waste produced. These countries, partly as a result of their corporatist political culture, have achieved consensus and co-operation between government and capitalist firms on the appropriate measures. Most obvious in this regard has been the creation of the Dutch National Environment Policy Plan, adopted in 1989.

In part this has also been the result of the relative willingness of capitalist enterprises in these countries to take a longer-term view of what will lead to economic viability than the annual statement of profit provides. Hence, in Germany the 'precautionary principle' has now gained widespread acceptance, namely that the absence of scientific certainty as to the nature and degree of an environmental threat does not justify doing nothing to tackle it. This is an exercise in foresighted prudence of a kind often held to be incompatible with the short-termism of capitalist practice.

However, as Dryzek also points out, the securing of a profit has not disappeared as the main capitalist aim. Rather, ecological modernisation is held to be good for business. He cites five main advantages for capitalist firms in climbing aboard this bandwagon. Less environmentally harmful forms of production are less inefficient; tackling problems at an early stage obviates worse costs later; employees are happier and work better in good environments; there is money to be made in selling goods and services with 'Green' credentials to an increasingly environment-conscious market; profits can be made from producing and selling products designed to reduce or prevent pollution. One might summarise this by saying that in these areas the conventional methods of accounting, attending solely to factors which can be given the usual financial expression, point capitalist corporations in a more environmentally beneficent direction.

Several important points for ecologism emerge from this phenomenon. First, it provides an example of capitalist enterprises apparently being willing to accept that governments have a legitimate role in trying to alter the organisation of capitalist firms in the name of the wider public good. Secondly, it shows that the long-term view is not hopelessly but of phase with capitalist concerns. Thirdly, it shows that the holistic, systems-oriented approach of ecology, and ecologism, concerning interactions between humanity and the rest of nature can be accepted by both capitalist enterprises and governments. Finally, it shows that even if the replacement of the market system by a fully planned economy is no longer a viable alternative, there is scope for planning in the sphere of economy-environment relations. The Dutch National Environment Policy Plan set environmental quality targets and a timescale within which they were to be achieved, based on a theoretical understanding of the processes involved.

All of these elements suggest that ecological modernisation is something which at least reveals the potential for a form of political economy within which ecological justice can be found a place—although, of course, this is still a long way off. The limitation of the scope of democracy represented by the guardianship institutions also looks less at variance with the capitalist system if the latter is itself willing to accept some legitimate government intervention with respect to its often trumpeted 'freedoms'.

However, ecological modernisation looks at the moment like a game for rich, successful capitalist countries who are willing to countenance some technological tinkering with their economic processes to achieve some clear-cut, if limited, environmental improvements within their borders, provided that profits and competitiveness are not compromised. But none of this may be affordable elsewhere, and the environmental injustices involved in the export of toxic waste, the expropriation of raw materials at low rates of remuneration from the Third World and so forth, may mean that this phenomenon is largely window-dressing.

Nevertheless, as Dryzek notes, these developments may open the door to a much more radical ecological restructuring of capitalism. Following Christoff and Hajer, Dryzek moots a development of ecological modernisation along the lines of the institutionalisation of democratic participation in the environmental planning of whole societies in a way which also encompasses the international dimension. At this point, then, in its turning to the need for wider democratic decision-making, the discussion of ecological modernisation joins up with the discussion of sustainable development in the last section. In the longer term, Dryzek suggests, this process might even lead to the 'exit from industrial society'. This might emerge as a result of the effect of what Beck has dubbed 'the risk society', which is one which generates severe environmental risks with which it has no effective way to deal.

However, we are still clearly embedded in an industrial society which, under the impulse of capitalism, is spreading to newer areas of the planet with, as yet, no serious impediment. It is worth noting at this point the features of capitalism which, in spite of the promise of ecological modernisation, make it problematical from the point of view of ecologism.

Capitalism remains resolutely human-chauvinist in its orientation. Human welfare and/or preference-satisfaction is the sole ethically important demand which it can admit. Thus, environmental issues are couched either in terms of prudence (what is instrumental for human welfare) or in human 'quality of life' terms. The idea that the non-human may possess moral considerability which circumscribes human economic activity is one which capitalism as such cannot encompass. This is because capitalism is designed only to respond to human preferences expressed in the marketplace. Although individuals who are capitalists are capable of wider moral concerns than this, *qua* capitalists they can only be concerned about market-expressed preferences.

However, capitalism's responsiveness even to human preferences is contingent on their being expressed in the marketplace. It cannot respond directly to human preferences not so expressed. This in turn means that, for capitalism to respond to non-human preferences and needs, the latter must first be reformulated as human preferences (for example, that certain species should continue to exist) and these must then be made manifest to capitalist enterprises by either or both of consumer preferences (for green products/services) and politically created regulations and 'market' mechanisms (taxes and subsidies). Hence, capitalism *qua* capitalism has no intrinsic tendency to show any responsiveness to non-human interests (or to many human interests either).

By contrast, the profit motive is central to capitalism. This fact alone makes the ideas of 'stakeholding' and 'ethical investment' hard to embed within capitalism, for they have the clear tendency to compete with that motive, and it is a motive which has to be given total allegiance by capitalists, on pain of failing in their capitalist enterprise. As we suggested above, proposals put to capitalists for greater environmental responsibility have to be shown to be commercially viable if they are to be given a hearing.

Further, if consumer preferences and government regulatory and 'market' activity threaten capitalist firms' profitability the latter will have a very strong tendency to try to mould or manipulate the consumer preferences and to influence politically the governmental actions so that opportunities for profit are not diminished.

The profit motive is not eliminable from capitalism, nor is the commitment to economic growth. Jackson has given a clear explanation of how these two phenomena are connected via the key ingredient of capitalist financial markets, namely, the charging of interest on the loans necessary for capital investment. Roughly, there can be no investment without loans, no loans without interest being charged, no payment of interest without the making of a profit and no maintenance of production (and thus securing of a livelihood) without further investment to be paid for by further loans, more interest and increased output and/or lower costs to repay them, all of which adds up to economic growth. The importance of this argument is that it points to forces within the structure of capitalism to explain why growth is an ineradicable feature of it. It does not rest on an analysis of capitalists as greedy or avaricious, even if some are and are attracted to a capitalist life because they are. Thus the cure for growth fixation will not come about as the result of making sure that capitalists are not subject to such vices.

An important threat to capitalist profitability comes, of course, from competitors in the marketplace. The competitiveness is crucial in stimulating efficiency, technological innovation and creating consumer power, which are held to be key benefits of the market system, heightened in its capitalist form by the private ownership of the means of production. But this need to compete is what provides a standing temptation to capitalism to cut corners, environmental or otherwise, to maintain or gain a competitive advantage.

If capitalism is necessarily committed to profitability, and if profitability requires growth in output, then capitalism is necessarily committed to growth. There are, as we have seen, ways of making economic growth less materially intensive, polluting and wasteful, but the impact on the environment in terms of use of resources and pollution cannot be reduced to zero. With constant or increasing rates of growth at some point the growth curve starts to rise steeply (the well-known phenomenon of exponential growth). We can defer this by reducing the rate of growth, but unless we can keep reducing the latter without limit, at some point the steep rise must kick in. Hence, the best we can do with capitalism is to reduce its environmental impact for a time until, as Jackson suggests, we find a different political economy to practice.

During this period we may hope that the one environmentally beneficial development internal to capitalism will continue to pay dividends, namely the discovery that making production less polluting and more parsimonious in use of material and energy cuts costs and aids profitability. But the points noted above entail that the main forces keeping capitalism pointing in a greenish direction must come from outside it—consumer pressure and governmental regulation and market activity.

Ecologism should conclude, therefore, that while much can be done immediately to make capitalism more environmentally benign, and economic growth, capitalist or otherwise, will be needed for some time to produce improvements in human well-being in the poorest countries, we do need to look beyond growth-oriented economic systems in general, and capitalist ones in particular, if we are to have a political economy which secures both human well-being and allows for the moral consider ability of the non-human.

What would such a system be like? We can say at this stage only that it would be one in which the material needs of a stable human population would be met in a manner which minimised material throughput and human beings' primary concern would switch from the accumulation of material wealth to non-material needs, such as affection, creativity, the contemplation of beauty and the pleasures of human intercourse (presumably themselves satisfied on a largely non-material basis).

Economic democracy

Let us finally survey the possibilities and problems, from the point of view of ecologism, of democratizing the internal structure of capitalist firms. This possibility, called 'economic democracy', has recently been discussed by Archer. The case for such democracy is derived by Archer from the basic value to be attributed to autonomy for individual human beings, understood as making one's own choices about how to act. The case for autonomy in joint decision-making leads to the case for democracy in general and to two specific principles:

(1) The 'all affected principle' — All individuals whose ability to make choices and act upon them is affected by the decisions of an association (such as a capitalist firm) should share control over the process by which those decisions are made.

(2) The 'all subjected principle' — All individuals who are subjected to the authority of an association should have direct control over the decisions of that association.

Principle (1) applies to the 'stakeholders' in a capitalist enterprise, comprising employees, consumers, shareholders, suppliers of raw materials and producer goods, financial institutions and local/non-local residents whose lives are affected by the firm's externalities. Such stakeholders' control has to be indirect, that is does not involve participating in the internal decisions of the firm. The two indirect methods of control available to stakeholders are 'exit', or removing oneself from relations with the firm as worker, consumer, shareholder and so on, and government regulation, especially where no exit is available as a means of exercising control.

Principle (2) applies to the employees who are, under capitalism, subject to the authority of the firm which employs them. That is, *qua* employees their actions are governed by the choices of the firm's managers, not by their own choices. It is with respect to them that Principle (2) enjoins the creation of direct decision-making within the enterprise. To shareholders would remain the exit strategy of control which is, Archer claims, what in most cases is the main instrument of shareholder control even under present circumstances.

The question of whether such decision-making would impair the economic efficiency of capitalist firms is, Archer argues, an open question. He notes, however, that if it could be conclusively demonstrated that economic democracy was bound to be much less efficient than present forms of capitalism then its acceptability would have to be seriously questioned. This is intelligible on the premise of autonomy, for an economy which was operating inefficiently would itself be detrimental to the exercise of human autonomy by depriving people of some possibilities of choice. However, he notes that the question of the viability of a system is value-laden and depends upon what we demand of it. If economic efficiency is not our sole goal, then we may reasonably accept a less efficient economic system in order to achieve some other important goal.

From the point of view of ecologism, the autonomy-based case for economic democracy is an attractive one, as already noted. However, one element in Archer's analysis is obviously open to criticism from this point of view, for it can be said that missing from the list of stakeholders, or those individuals affected by the decisions of associations, are non-human living beings whose well-being may well be adversely affected by associations' activities. For these stakeholders, exit is not possible as a means of control, because they are not in conscious relationships with human associations, and so cannot employ exit to assert such control. Some animals may be able to flee adverse effects, but, even if this, for some reason, inflicts a penalty on the firm, it still does not amount to control. Nor, as non-participants in politics, can they directly obtain government regulation to protect their interests.

Of course, the benefits of both exit and regulation are available to them indirectly, provided that human consumers and shareholders withdraw from relations with firms which ignore requirements of ecological justice, and elected politicians enact regulations with the support of concerned citizens to enforce the requirements of such justice.

What this highlights is that, from the point of view of ecologism, economic democracy's prime justification will be on the basis of human autonomy, but that it will require the practices of economic democracy to be subject to external forms of control, especially from the guardianship institutions at all levels, aiming to ensure that ecological justice requirements are met. There is nothing inherent in the idea of any kind of democracy which guarantees that ecological (or environmental) justice is respected. This applies also to economic democracy.

Having surveyed the reasons why ecologism should be prepared to offer an extensive critique of the currently dominant form of political economy, and having surveyed the reasons for both optimism and caution concerning the possibility of transforming it in ecologically benign directions, we need finally to address arguments coming from the other end of the current debate. One recent Green discussion of this kind involves a thorough-going scepticism directed towards the rhetoric of sustainable development and attempts to produce a version of political economy which is very unlike any form of contemporary capitalism. We should first, however, consider an ingenious alternative to capitalism emanating from the socialist, rather than the Green, camp and assess it from the point of view of ecologism.

❑❑❑

4

Alternatives to 'Greened' Capitalism: Market Socialism and Global Ecology

Market socialism

THE MOST THOROUGH RECENT case for market socialism is that put forward by David Miller. He attempted to find a viable socialist political economy in an age in which the most obvious alternative to liberal capitalism, the centrally planned economies of state socialism, seemed to have been comprehensively shown to be non-viable, and in which the ascendancy of increasingly radical forms of economic liberalism appeared to be unchallengeable. The theoretical arguments of Hayek, which establish what appears to be a conclusive epistemological case against the wholesale replacement of markets with central planning, are particularly important in this regard. In a nutshell these arguments amount to the claim that the knowledge needed by central planners to determine the productive and distributive activities of a large-scale complex economy is not available to them even in principle, for it concerns the needs and desires of millions of individuals, which only those individuals can know at any given moment.

Miller accordingly tried to take account of the historical and theoretical reasons for saying that a market system of some kind is indispensable to a viable economy, while showing, for the benefit of socialists, that such a system could meet the requirements of distributive justice and the need to avoid, or at least minimise, the evils for long held to be central to capitalism, namely exploitation and alienation. Against economic liberals it had to be shown that an alternative market system to that of capitalism could be coherently described which could possess much of the dynamism of capitalism.

Although ecologism has an interest in these matters from the point of view of human well-being alone, deriving from its basic concern with the conditions of human flourishing, it also needs to be satisfied that any alternative to capitalist markets is better able than capitalism to satisfy the requirements of ecological justice. This is an issue which, not surprisingly, Miller did not significantly address, given the aims just outlined. Can we supplement his discussion in this area, or must we conclude that on this matter market socialism and market capitalism are indistinguishable?

The 'pure' model of market socialism with which Miller is concerned is outlined as follows:

> All productive enterprises are constituted as workers' co-operatives, leasing their capital from an outside investment agency. Each enterprise makes its own decisions about products, methods of production, prices, etc., and competes for custom in the market. Net profits form a pool out of which incomes are paid. Each enterprise is democratically controlled by those who work for it, and among the issues they must decide is how to distribute income within the co-operative.

The reasons for evaluating this version of market socialism positively from the point of view of ecologism are that it does attempt to build in procedures of economic decision-making which *prima facie* attend to the autonomy needs of participant workers, and allows the wider society a stake in the productive activities of its members via the control exercised by a State Investment Agency (or Agencies) of the allocation of resources for investment. Further, Miller's case for saying that market socialism reduces the possibility of exploitation and alienation as compared with capitalist markets has some plausibility, and a concern for human flourishing has to involve a concern to reduce these phenomenal.

However, a case can be made from the point of view of ecologism for saying that the avoidance of alienation is not necessarily the most important consideration in the working of economic systems. If the preservation of the conditions of existence and flourishing of the non-human requires the toleration of human economic practices which are to some degree alienating, then that is arguably a price that ecological justice rightly exacts. Much will depend on the exact circumstances in a given instance as to whether this view is justifiable from the point of view of ecologism.

The reasons for concern about whether or not market socialism is better able than capitalism to fulfil the requirements of ecological justice stem, obviously, from its market character. Worker co-operatives are expected to compete and to make a profit. They are not to be rescued from bankruptcy if their productive activities fail in the marketplace, for the exercise of consumer choice (exit, again) as an essential method of determining allocation of productive resources is retained from market capitalism and is what enables the State Investment Agency to possess the requisite knowledge of where society would prefer investment to go.

Will worker co-operatives be willing to alter or amend their productive practices in order to meet the requirement of ecological justice, rather than to meet the purely anthropocentric aim of keeping human beings in employment? There is no a *priori* reason to suppose so, for such justice requirements are extrinsic to productive activities and thus can be ignored without direct jeopardy to the well-being of a given group of workers. The laying-off of productive workers, by contrast, is an inherent possibility in any kind of market economic system we choose to contemplate, and thus must be entertained as a possibility by those operating the system (such considerations also apply to the working of economic democracy within a capitalist system, of course).

Thus, there seems to be no reason drawn from the very structure of the market socialist system, at least as described by Miller, to suppose that the productive units themselves will have an inbuilt tendency to attend to the requirements of ecological justice. They may be more easily made to do so by the coercive power of the state than is possible under capitalism. But the proposal to use the State Investment Agency for this purpose runs up against bureaucratic politics. A guardianship body within a market socialist economic system may well find that it has a difficult job of ensuring that a powerful agency devoted to seeking profit-making opportunities within a market system pays much more than lipservice to the requirements of ecological justice. These may be no greater difficulties than are encountered in capitalism, where powerful private corporations also offer such resistance, but arguably they may be no less.

Market socialism, then, does not appear from the point of view of ecologism to be the panacea for the environmental ills perpetrated by the capitalist and state socialist systems. It is the promised improvement in human well-being which makes it attractive to ecologism, on the assumption that it does amount to a system with the ability to deliver the goods necessary for human well-being with a much reduced tendency to alienation and exploitation. But its ability to offer a significantly superior performance in the area of ecological justice is clearly open to question.

Global ecology

The most radical critique of the possibility of sustainable development, whether as articulated by environmental (neo-classical) or ecological economics, has been put forward by supporters of the view which has been dubbed 'global ecology', after the collection of articles of that name edited by Sachs. This view represents a relaunch of the environmental critiques prominent in the 1960s and 1970s which, according to the view, have been subverted and rendered harmless to the forces of capitalist-motored economic growth by the arrival of the rhetoric of sustainable development subsequent to the Brundtland report.

The critique of sustainable development embodied in this view may be expressed in a number of negative and positive positions. The following summary is based on the ideas propounded in the Sachs volume.

(1) On the negative side, global ecology first supports the abandonment of development as the aim of human economic systems, primarily because of its renewed emphasis on the limits to growth inherent in a finite planet. The idea of sustainable development is analysed as a rhetorical device to permit the hegemony of global capital and enhance the power of 'global' politicians, bureaucrats and

'experts', all of whom are drawn from the 'advanced' world of capitalism. Even the distinction between growth and development made by those who are critical of Western notions of economic growth, such as Herman Daly, is rejected as an impossible one. The obverse of this anti-growth position is the emphasis on the need to foster modes of political economy which are simpler, more sustainable in a fully ecological sense, less or anti-consumerist and focused on usevalue rather than exchange value.

(2) The processes of international environmental regime-building associated with such events as the Stockholm and Rio conferences are analysed as at best useless and as more likely than not to be positively harmful to environmental concerns. Their sole outcome has been to ensconce an international, capitalist-based, Western-dominated elite in a position of managerialist power, offering panaceas which involve the technology and scientific knowledge of Western experts and corporations and which do nothing significant to deal with the 'global' environmental issues of global warming and the rest which are their ostensible concern. Their main outcome has been to draw some, mainly Western-based, Non-Governmental Organisations (NGOs) into the embrace of Western states and corporations so as to give a superficial respectability to the proceedings.

(3) Global ecology claims that the idea of 'development' embodied in 'sustainable development' remains that based on the view of a single economic path into the future, namely the one already taken by Western liberal capitalist societies, which the 'underdeveloped' regions of the world are to be encouraged to follow, albeit now in some mysteriously 'sustainable' manner. In fact, it will be the West which keeps the 'development' side and the Third World which will supply the 'sustainability' in the form of providing the sinks for Western wastes and the curtailment of its own economic growth in order to compensate for the environmentally harmful practices of the west.

(4) The one-sidedness of 'sustainable development' rhetoric is disguised by its employment of the metaphor of 'one Earth' and of its continual reference to an undifferentiated 'humanity' said to be faced with a single set of problems which all human beings have to co-operate together to solve. The fact that the problems have been overwhelmingly caused by the West and that only if it changes its ways quickly and radically is there any hope of dealing with them is conveniently hidden by this rhetoric. The case for development is even bolstered by the argument that the main threat of environmental destruction now comes from the poverty of the Third World, and that the poor of those countries need to move rapidly further along the Western-style growth path in order to be able to deal properly with 'their' environmental problems.

(5) Global ecology argues that in the past the nature of Western 'development assistance' has usually been positively harmful, and certainly inappropriate, to the needs of the ordinary citizens of the recipient countries. Similarly, the kinds of technological fixes tied into such post-Rio developments as the World Bank's Global Environmental Facility represent the imposition of inappropriate solutions concocted at the global, state-centred level of international managerialism, by scientific and other Western-trained 'experts'. The net beneficiaries of all this remain the usual ones — Western Multi-National Corporations (MNCs) and Third World elites.

(6) A specific target of criticism is the practice, inherent in the process of capital accumulation as traditionally conceived in capitalist systems, of expropriation and conversion into state or private property of the remaining commons, especially those which remain in the Third World countries. The positive view of the role of commons is set out below.

(7) The critique of the pernicious effect of Western science and expertise is extended to encompass even the science of ecology. In its most recent varieties, it is said to have been unable to give any firm guidance on the issue of what

human beings can do to nature without disrupting and destroying natural systems. Scientific ecology is now claimed to have abandoned its earlier certainties about the existence of stable patterns and climaxes in the natural world and to be portraying the latter as devoid of any inherent balance. Where there is no equilibrium or balance there can be no disequilibrium or imbalance, Thus, the only scientific discipline that looked capable of offering some reasoned check to the harmful expansivism of science-based technology and economics in general has turned into a paper tiger.

(8) As is apparent from the earlier points, global ecology rejects or is profoundly suspicious of universalist discourse, whether scientific or moral-political, of the kind used to underpin the practices of international debate on environmental and related matters, and to formulate the 'solutions' to 'global' problems in international forums. It thus takes up again some of the earlier environmentalist cudgels against the Enlightenment inheritance of universalist conceptions of reason. It counterposes to this Enlightenment version of reason the ideas of immanence, diversity—of languages, forms of knowledge and cultures—and emphasises the impossibility of blending these together into a single universal discourse.

The positive conception of an ecologically sustainable form of political economy, which takes proper account of the moral considerability of the non-human, has inevitably been alluded to in the above critique. It involves a version of the communitarian outlook which we have already encountered in previous chapters, and may be summarized as follows.

The ideal form of human socio-economic organisation for global ecology is closely modelled on the traditional, commune-sized and largely agrarian form still to be found in parts of India. Such communes are viewed as having close interconnections with the specific natural environments within which they are located. These environments provide the resources, of course, to sustain the lives of communities, but also have a sizeable impact on the development of the cultures of the communities which interact with them. The members of such communities develop traditions, customs and other practices which enable the community to preserve the ecosystem within which they exist and thereby to preserve the biodiversity which is bound up with it.

The knowledge which such communities possess is not, of course, scientific, theoretical knowledge of the kind which is so celebrated in the West, but is a practice-based knowledge, developed over long periods, and highly attuned to the needs of the complex ecosystem which the people inhabit. Such knowledge is a communal possession, embedded in the culture of the specific commune. Hence, there is a close connection between the preservation of non-human diversity, or biodiversity, and human cultural diversity. Contrariwise, the flattening-out of human cultural diversity and the reduction in biodiversity are also interlinked.

The impact of specific ecosystems on specific human cultures involves the interlinking of practical requirements, such as the need for periods when the land is left fallow, and religious practices and mythic constructs which interweave the practical with the symbolic, aesthetic and social—as in certain festivals and ceremonies. The specific ecosystems within which such traditions develop and have their point furnish the natural patterns, symbols and experiences which give the traditions their concrete form as well as their practical purpose. Western observers, imbued with rationalistic, theory-laden preconceptions, will tend to view such events as outmoded primitivism, picturesque but doomed to be swept away, or perhaps preserved for the benefit of tourism, once 'progress' and 'development' have been set in train.

The specific mode of political economy which is inherent in the traditional commune is the commons. The resources upon which the commune depends are viewed as communally owned and access to them is regulated by the commune itself, via a series of traditional rules and practices accepted by all the members. This is emphatically not the same as the commons famously analysed by Garret Hardin, which is in reality an 'open access' regime, in which access to the resources is completely uncontrolled. Commons properly so called are well adapted to the preservation and biodiversity of the ecosystem involved. Thus, they do not need the imposition of a state-based mode of coercion in order to prevent self-interested individuals from being trapped in a prisoner's dilemma, nor do they need to be enclosed and constituted as private property so as to be 'looked alter properly' and 'developed rationally' in the manner beloved by liberal capitalism. Both of these favourite remedies of the West lead instead to the ecologically harmful and culturally catastrophic imposition of monocultures, destroying species and depriving people of their history, culture and identity.

The aim of global ecology, therefore, is to seek to defend the commons-based communities which still exist, albeit in dwindling numbers, against the evils of enclosure and 'development'. Since the whole world is now divided up into states, this means in practical political terms seeking to protect and enhance as far as possible the autonomy of communes from central government, dominated by Western-oriented elites, and international organisations. It also means resisting the favourite Western state-directed solution for the protection of biodiversity, namely the creation of national parks and conservation areas from which the traditional peoples are excluded and/or with respect to which they are prevented from practising their traditional techniques of environmental management.

This is not to say that all such communes are perfect enclaves within a wicked world. However, they are held to be a sounder basis for the preservation of the non-human world and of human well-being than are the consumerist mass societies of the liberal capitalist world. The main proposals put forward by global ecologists for the improvement of traditional communes are: their thorough-going democratisation; their voluntary reorganisation along more explicitly bioregional/ecosystem lines, overcoming the arbitrary frontiers created by state bureaucrats; the development of new forms of intercommunal relationships, democratically developed and aiming to provide communes with mutual aid to fill gaps in each other's natural resource base.

The standard of living attainable in such communes is expected to be one which achieves a reasonable sufficiency for human beings, but which nevertheless will be frugal in comparison with the commodity-glutted West. The gains achievable by communes in the cultural and spiritual richness of life associated with the preservation of a diverse non-human world, humanity of scale and the experience of rootedness, are expected to furnish more than ample compensation for the simplicity of material existence within them.

Let us now consider some of the difficulties with global ecology as outlined above. First, one may properly wonder how accurate are the views expressed by devotees of global ecology concerning the ecologically beneficial nature of traditional agrarian communes. There is certainly evidence that even in precolonial times societies organized on such a basis have played their part in major extinctions and destruction of habitats — consider the extermination of flightless birds by the Maoris in New Zealand, for example. There may be a fallacy lurking in the global ecologists, celebration of the closeness of ecological-cultural interconnections in traditional societies. This is to suppose that because human culture and non-human nature are closely interwoven in such societies all of the non-human will receive care. It is at least possible that such interconnections are often rather partial, with some aspects of the non-human natural world being accorded concern, but others being neglected or attacked, particularly if they have no obvious economic or cultural value to the people involved.

To avoid the charge that global ecology is peddling the myth of the golden age, it would be as well for its supporters to admit the probable deficiencies of all human societies in terms of their ecological soundness. The case for communes and commons is one which need only refer to a marked difference in degree of ecological harmfulness between traditional and modern systems. Differences of degree are often of vital importance.

Secondly, one may well wonder whether all development, even that associated with the Western concept, is to be shunned, especially if one is concerned to promote the conditions for well-being and flourishing of the human species. Traditional communes in agrarian societies such as precolonial India may have been rather more agreeable places than they have become in a modern world in which the basis of their subsistence has been undermined by remote and bungling bureaucracy and/or market-based economic pressures. However, they did face chronic problems of malnutrition, famine, high infant mortality, short life-expectancy, disease, squalor, drudgery and monotony of diet as well as being characterized by such social pathologies as sexism, ignorance, fear and oppressiveness. At any rate, many late twentieth-century inhabitants of these traditional communities do conceive of the latter as prey to such deficiencies. They are also aware of the Western model of development as promising to remedy some or all of these deficiencies, even if at the expense of destroying ancient traditions and despoiling ecosystems.

It looks, therefore, as if an alternative, ecologically sustainable form of community based on commons must be prepared to argue for the necessity for schools, health care and appropriate technology to relieve the tedium and drudgery of traditional communal life. Global ecology aims also to create democratic, participatory and tolerant communities which, while cherishing their own traditions, are open to those of other communities and are not closed in upon themselves. It is hard to see why these desiderata should not be spoken of as development or progress, and thus it seems clear that there has to be room for at least some developmental purposes to be admitted, and ones which are clearly part of the universalist, Western version of the concept too.

In any case, the elites of non-Western countries are clearly committed to 'catching up' with the West — a project understood almost completely in the terms established by liberal capitalism. There may be some residual concern about the loss of distinctive cultural traits in the process, but not, apparently, to the degree necessary to lead such elites to try to put a halt to the 'catching-up' process. Among the non-elite of countries such as India, there may be greater misgivings about the impact of this process on traditional cultural values. However, many of the ordinary people based in traditional communes have already found themselves having to leave the land for the cities in order to try to acquire a wage necessary to purchase what they once were able to grow for themselves, as more and more agricultural land is turned over to lucrative cash crops and monoculture food crops, sold in the national marketplace by the new farming elites.

In the cities, of course, country people are relentlessly exposed to Western consumerist, individualistic values and notions of progress, assuming they have not already encountered them in the countryside. In countries such as India and China, where entrepreneurial values and skills have always had a part to play in the life of the society, the new commercially minded groups experience the emphasis of their governments on economic growth and commercial success as a liberation from ancient shibboleths and failures, not as something to be resisted.

The point of these reflections is, obviously, that global ecology's emphasis on the importance of communities and communes, even in those societies where they still may be found, may have arrived too late to be capable of strengthening traditional forms against the undermining tendencies of powerful economic and cultural forces. However, if ecologism is correct in its view that Western models of development do portend ecological disaster, then this may be sufficient to turn what at present appears to be an almost unstoppable force into a passing fashion. In countries such as India, the ideas of Gandhi may eventually win the battle with the ideas of Adam Smith.

A third problem with the global ecologists' diagnosis is the lack of any very convincing prescriptions for what should happen in those societies in which it is virtually impossible any longer to find examples of commons and traditional communes. It is at least intelligible to propose the attempt to preserve existing communes and commons, but how does one re-establish such things when they have gone almost completely—apart from the attempts of environmental or religious groups to create communes within the interstices of Western liberal democracies? Of course, from the point of view of a concern for biodiversity, the global ecologists' strategy at least has the merit that the societies within which traditional communes are still to be found are the very ones which are crucial for maintaining biodiversity. Hence, preserving traditional modes of organisation in these may appear to be both practicable and vital. The 'developed' world has already foregone many of its ecological riches in pursuit of its materialist purposes, so much less importance may attach to trying to recapture the ecologically beneficial mode of political economy in such societies.

However, the problem with this is, of course, that the adverse effects on the world environment associated with global warming, ozone depletion and so on are largely the responsibility of such countries' activities. Also, these countries, with the dynamic capitalist market system at their heart, are inherently expansionist and dominate the world economic system, exploiting the traditional commune-based societies for their raw materials and sink-providing capacities. Clearly, if they remain on their present trajectory there is no real prospect of erecting a great wall of China against their ecologically disastrous activities. They must be greened in some way, and global ecology needs an account of how that can be done effectively and quickly.

A related difficulty with the global ecologists' analysis is that it is silent on the question of how the ideas of the commons and communal organisation may be applied to urban existence, where an increasing proportion of the human population lives, in both the 'developed' and the 'underdeveloped' regions of the world. Of course, the idea of 'commons' regulated by a 'commune' is not totally inapplicable to city life—medieval guilds may provide an example. However, the modern city is one where productive resources are overwhelmingly in private hands and where most people live individualised modes of existence as wage earners.

Further, the attraction of the agrarian commune as a model of economic existence is that it contains a clear conception of the interpenetration of ecology and culture, for on that model specific communes relate to specific ecosystems, which they both manage and are shaped by. However, modern city life is the paradigm of a mode of existence in which human beings are cut off almost entirely from daily interaction with natural ecosystems. It is a strength of the global ecologists' analysis that this explains some of the anomie and cultural uniformity of the city dweller's life, but it is a weakness that it has nothing to say about how the milieu within which a majority of the planet's human inhabitants will shortly be living should be altered in order to develop the kind of cultural/ecological knowledge upon which the future environmental health of the planet depends.

It would seem an implication of the global ecologists' analysis that cities, especially the huge conurbations within which many human beings now live, should ideally be diminished in size, in terms of their populations, their geographical extent, the amount of energy and resources they consume, and the amount of consumer goods and associated waste which they produce. This may involve somehow inducing a voluntary return of people to a countryside organised along commune lines—the example of enforced rustication of people for ideological reasons provided by Cambodia under Pol Pot causes a shudder at this point. Conceivably the aims of global ecology might be met by the penetration of city life by agrarian activities, such as city farms, and the spread of large areas of reclaimed habitat within city limits. Without some sort of an account of what is to be done about cities the global ecologists' analysis is bereft of a strategy for dealing with a massive and accelerating human development which promises to swamp any gains for ecological well-being made in the remaining agrarian parts of the world not yet taken over by city-based agribusiness.

Finally, there are some objections one may have of a more theoretical kind to the analysis provided by global ecology. First, the analysis endorses a view of human culture, encompassing values and conceptions as well as artifacts, which emphasises boundedness and non-translatability, which sits ill with the use of such universalist concepts as democracy, justice and autonomy in its prescriptions for the relations between communes and states. In its recognition that actual examples of communes are far from perfect in many ways—they are subject to forms of oppression and authoritarianism as well as sometimes being properly subject to adverse criticism of the precise nature of their relations to the natural world — global ecologism appears to adopt a transcendent standpoint which its substantive theory of language and culture rules out as impossible.

In its own critique of the relations between liberal capitalism and Third World cultures, the former are depicted as straightforwardly dominating and exploiting the latter in terms of concepts of domination and exploitation which are intended to have cross-cultural applicability. That is, the diagnosis of exploitation is one which both members of the exploiting group and those exploited are expected to recognise as correct, and as correct in the same way. Global ecologists, whatever their native culture, are expected to concur with this diagnosis, which accordingly appears to be culture-transcendent. But if certain universalist concepts are admitted as legitimate it is not clear why others should be rejected. Their universalism cannot itself be an objection.

In any case, there is a worrying reification apparently at work in the global ecologists' account of culture-ecosystem relations. Both specific ecosystems and the specific cultures which they support and with which they interact are depicted as having a fixed boundedness which is inapplicable to either phenomenon. It is of course possible to individuate cultures and ecosystems for certain purposes and within certain rough limits, but one great theme which ecologism has correctly emphasised is the interconnectedness of phenomena, whether human or non-human. Specific ecosystems and human cultures are interconnected across time and at a time. Boundaries are drawn for certain purposes and redrawn when those purposes change.

It is possible that the emphasis on diversity and distinctiveness is not intended by global ecologism to rule out such interconnectedness, and that the aim is simply to redress the balance somewhat in a discussion which has become too 'global' in its orientation. But there is clearly the danger in taking this line that the non-human world will not be treated as a unified whole when in certain respects it is important so to treat it, and that the interconnections between human beings, moral and cultural, will be neglected or downplayed to the detriment of both culture and morality.

Finally, we must consider the rather surprising rejection of scientific ecology as any sort of a guide to how human beings should interact with the natural world so as to further their own well-being without significantly reducing the well-being of non-humans. It may be true that the most recent theorising in the science of ecology has jettisoned any idea of the natural world as exemplifying patterns of stability and equilibrium which give human beings a benchmark against which to measure their impact on the non-human system. It does not follow from this that nothing human beings do can adversely affect the non-human world, and that there are no ecological reasons for objecting to at least some human activity carried out in the name of development. After all, human activity has demonstrably brought about the extinction of some species, has altered habitats to the detriment of both humans and non-humans, and has even altered large parts of the purely physical environment of the planet, such as the ozone layer, weather systems, wetlands and water courses.

In deciding whether or not these are causes for congratulation or regret it may no longer be possible to refer the changes to some theoretically ascertainable situation of equilibrium which non-human nature would have attained in the absence of human activity. But it is clear that key ecological concepts, such as those of ecosystem, negative and positive feedback, keystone species, adaptedness, niches and so forth will need to be employed in order to understand what has happened, what the implications for the future are, and what the effects are likely to be on the well-being and even existence of life forms. Consider, to take only one example, the crucial importance for the assessment of extinction rates, of the formula for determining numbers of species in island biogeography. Hence, the scientific theorising embodied in ecology is something which cannot reasonably be rejected in favour of the practical knowledge inherent in the traditions of agrarian communes, valuable though that undoubtedly often is.

Now that we have reviewed the case made by global ecology and considered some of the difficulties which the view faces, we need to decide how much of it is defensible. The following points are of value in global ecology.

It is reasonable to be suspicious of the concept of 'globalism' as a conduit for Western, capitalist-based, technocratic managerialism which has a tendency, via international organisations and regimes, to impose pseudo- or non-solutions to an environmental crisis which, up to the late twentieth century, has largely been of Western making. The main beneficiaries of such 'solutions' are probably disproportionately Western-based corporations and parts of Western populations (understanding 'benefit' in traditional economic terms). It is correct to insist that the crisis is one in which the culprit is not 'humanity' and in which not all members of 'humanity' suffer equally, even with respect to such 'global' phenomena as climate change.

The obverse of this position also has a great deal to be said for it, namely that genuine solutions to environmental problems are usually best found at the local level, using local knowledge of specific ecosystems and engaging the interest and concern of people directly involved with them, rather than centrally imposed solutions, based on abstract theory, which involve the attempt to separate local people from the ecosystems in question.

The political implications of this are also valuable, namely the case which is made for decentralisation of decision-making and the greater autonomy of localities from the central state, particularly when coupled with the analysis of the self-regulatory nature of the commons as a mode of economic organisation in an agrarian milieu. The preservation of existing commons and traditional communes is certainly an important consideration in the attempt to see how human and non-human interests may be brought into some kind of harmony.

There is an interesting parallel here to the argument we noted at the start of this chapter which was put forward by Hayek in favour of market-based decision-making and against central planning of the economy. He argued that the knowledge of preferences upon which allocative decisions would have to be made can exist only in the minds of the individuals whose preferences are involved. Such knowledge cannot be gathered into one place to form the basis of a grandiose economic plan. Similarly, global ecology reminds us of the enormous multiplicity of ecosystems which underly scientific ecology's generalisations. This implies that the knowledge of the interactions which make up the actual biosphere is to be found only in the minds of people who are interacting with those systems at the micro level, and thus it is to them, rather than to a central body of scientifically informed experts, that we should look for solutions to serious environmental problems.

Such knowledge is inevitably partial, but has the great advantage that it is derived by methods of direct interaction which enable its possessors to track the specific changes which are ceaseless in any such system, and to accumulate an inductive understanding of what is likely to harm or benefit the local system. For this to work properly, of course, the people operating at this level need an incentive to engage in the effort to discover how their local ecosystem works, which is a further reason for emphasizing the importance of the commons as a mode of economic organisation.

This implies in turn that the proper role of the central state with respect to the implementation of this environmental knowledge, embedded in specific cultural formations, is to act as a facilitator rather than as an imposer of solutions. One valuable service it can perform is to protect the localities against those forces which threaten to disrupt the cultural formations involved and to provide environmental information not available to the specific localities as well as fostering their co-operative relations with neighboring communes.

However, these claims leave intact the case for the creation of a centralized 'guardianship' institution, to monitor changes in the structure and practices of communes in societies where they are in existence. Where they do not exist, and cannot easily be revived, the role of the guardianship institution remains vital, and it must perforce be one which is largely informed by the theoretical analyses of scientific ecology and allied disciplines.

The strength of the case for the commons as (at least potentially) an ecologically sound mode of political economy does support the project of searching for ways to protect commons and communes where they exist in capitalist-dominated societies and for seeking ways to extend their activities. But this, as we noted above, is going to be an extremely difficult task in a West oriented so wholeheartedly to individualism, markets, consumerism and technological innovation.

The general case against the rhetoric of development, sustainable or otherwise, is a valuable reminder of how radical modern environmental critiques are. For development and progress are the great Enlightenment aims, the hallmark of modernity, and in challenging these notions environmental critiques penetrate to the heart of the issue of what humanity should be aiming to achieve on this planet — chiefly by making us suspicious of any rhetoric which postulates that 'humanity' as such has any single aim at all. The renewed emphasis on limits to 'growth' and the justifiable skepticism about whether Western models of growth have been helpful to the well-being of Third World people and their environments are other valuable features of global ecology.

However, it is harder to accept that all that goes under the name of 'development' is equally harmful. Some changes are clearly necessary for the improvement of human flourishing in the Third World—and in parts of the liberal capitalist world too. These changes may be detachable from the model of development which supposes that the path followed by the West is the one true path which all must follow, but there will inevitably be some overlap with it. Health care, education, improved nutrition, basic infrastructure—transport and communications, sewerage and clean water supplies—all are essential to improving the well-being of human beings in even the most ecologically exemplary communes.

Arguably so too are some systems of social welfare to help in cases of old age and illness. The commune may well be able to provide welfare assistance on the basis of mutual aid and solidarity, but that will depend on how close to subsistence level any particular commune is. Communes may expect the individual family to be the prime provider of such welfare, in which case if there is poor health care within them, with high levels of infant mortality, we are faced with the phenomenon of people having large families in order to have someone to depend on in old age, one of the prime factors driving population increase in Third World countries.

It is hard to see how these elements of development are achievable without the generation of economic surpluses. These are needed to provide the capital accumulation necessary to finance these developments. The latter require not simply material capital for construction, but the human capital of medical, engineering, educational and financial skills and associated institutions. This suggests that a purely subsistence form of commune-based economy has to be rejected. But such rejection in turn implies a commitment to a certain amount of economic growth. The problem then is how to achieve this in an amount sufficient to finance the humanly necessary growth without putting into play forces, such as the creation of products for exchange value, the development of markets, specialization and consumer products, which lead on willy-nilly to the full-fledged Western model.

It is here that a distinction between a purely quantitative concept of growth and a qualitative concept of development of the kind supported by Daly has a point, and it is hard to see how a defensible ecological position can be developed without it, even if in practice it is hard to offer a clear and watertight specification of what it involves. It is, however, a valuable suggestion one might take from the localism inherent in global ecology that there should not be a single overarching concept of development, but that the content of the concept should be determined largely by localities on the basis of their particular cultural formations.

Part of what is likely to be involved in the qualitative concept of development produced in traditional communes is a concept of praxis which downplays the idea of personal autonomy and self-development present in the Western ideal of progress and which instead emphasises the satisfaction to be gained from working within a community, jointly managed with one's neighbours, in harmony with nature. However, it will be recollected that one respect in which global ecology seeks to modify many traditional communes is in the direction of greater democracy and participation, which implies that the ideal of personal autonomy, although downplayed in relation to the ideal of the autonomy of the commune, also has a place.

The upshot of all this is to view global ecology as offering some salutary counters to a concept of sustainable development which is in danger of simply acting as window-dressing for the Western growth mania which has brought us to our present predicament in the first place. Its emphasis on decentralisation, diversity of ecosystems and culture and of their interplay, and on the subordinate role to be played by states; its scepticism of Western technocracy and crypto-imperialism; its celebration and analysis of the ecological wisdom present in many traditional communes and its emphasis on the importance of distinguishing real commons from open access regimes and on the vital necessity to prevent the remaining ones from suffering expropriation in the name of a misconceived concept of economic development, are all valuable.

But it is clear that some of the enlightenment inheritance needs to be retained, even to make sense of some of global ecology's own prescriptions. The state remains important as a facilitator of commons and as a key regulator, with other states, of the open access systems of the planet. Some ecological phenomena cannot be dealt with by localized communes, but need some form of international co-operation. It may be misleading to refer to these as 'global', with the implication that they are not primarily the result of the activities of the Western liberal capitalist societies, but they nevertheless cannot be dealt with on a localized basis. Also the prospects for the creation of communes based on commons in the world of liberal capitalism do not seem very promising, so that the state has to adopt a much more active role in regulating and controlling environmentally harmful activities there.

The aim of ecologism with respect to the liberal capitalist world has to remain that of developing and gaining acceptance for a distinction between growth and development and of seeking to clarify the concept of sustainability in a way which does not simply allow the present show to carry on as before. In this regard, as we concluded at the end of the previous chapter, the concept of sustainability promulgated by ecological economics appears to be the most promising one. Its emphasis on the need to make economic decisions in a manner which involves democratic discussion and debate at least complements the democratic, participatory emphasis of global ecologism.

With respect to the Third World, some elements of growth, and the scientific and technological underpinning which that requires, remain vital to human well-being. Until life expectancy, child mortality, morbidity and illiteracy rates show dramatic improvement it is going to be very hard plausibly to maintain, even to their inhabitants, that traditional agrarian communes, ecologically estimable as they may be in other respects, are adequate to sustain human flourishing.

Although global ecologism expresses understandable skepticism about science and technology, and emphasises the merits of traditional forms of knowledge, it is clear that scientific knowledge, especially in the form of scientific ecology, remains of vital importance to an environmentally sound future.

What one might conclude is that global ecology aims to produce a greater evening-up in the conditions of the human inhabitants of the planet not by hastening the 'underdeveloped' parts of the world along the trail blazed by capitalist liberalism, but rather by moving both the West and the Third World more towards each other. The Third World should be developed without following the Western road, and the valuable aspects of the Third World societies — communalism, commons, closeness to the non-human world, greater humility in the face of nature — should be protected and developed within the West.

However, the upshot of this chapter is to leave ecologism with no clear alternative to seeking the transformation of liberal capitalist political economy. Since capitalism undoubtedly exists, is growing in scope and is attracting the allegiance of billions of new adherents, realism, if nothing else, requires that the direction in which ecologism presently must look is towards the extensive and rigorous scrutiny of ways in which to force ecological and environmental justice on to capitalist enterprises.

This is its distinctive position, rather than any specific proposals for a radically new economic system (although, as we have seen, it has reasons for supporting at least some of these). Clearly, this is a large and complex task, but it is of considerable help to ecologism that it knows what it wants to do in the sphere of the human economy, and why it wants to do it. If we are clear about that then the much vaunted human ingenuity may be turned in an environmentally benevolent direction and show us how to do it.

❑❑❑

5

Towards a Climate-Neutral World

Introduction

HUMAN-INDUCED CLIMATE CHANGE presents society with a long-term problem. The challenge is to find ways of reducing CO_2 emissions in the short term that also enable society to reach long-term goals. Over the last 20 years, it has become increasingly clear that human activities are adding to natural changes in the climate. The enhanced greenhouse effect caused by emissions of greenhouse gases related to human activities is a major factor in currently observed climate changes, as well as in climate change that is expected to occur in the coming centuries (IPCC, 2001a).

The long-term character of climate change (having an effect for up to several centuries) is one of the key features of the climate change problem. However, dealing with long-term problems is generally difficult for modern societies, in which political agendas and practices tend to focus on the next shareholders' meeting or election rather than on the needs of future generations. If we are to minimize the human contribution to climate change, society will have to move towards a 'future of low greenhouse gas emissions' or in a 'climate-neutral direction'.

If society wants to realize what we refer to in this book as a 'climate-neutral society' (*i.e.*, a society that minimizes its negative impact on the climate system and its contribution to climate change), substantial reductions of up to 60–80 per cent (compared to the 1990 levels) in the greenhouse gas emissions of industrialized countries will have to be achieved before the end of the century, For Western societies, this implies a major shift in current and expected trends in emissions—emissions that are expected to rise in the absence of stringent sustainability and/or climate policies (IPCC, 2000; UNDP, 2000). Achieving such drastic reductions in emissions will be no mean task since societies—particularly Western societies—rely heavily on the availability of relatively cheap fossil fuels and the idea that the atmosphere is a free common good into which greenhouse gases can be emitted. As the Intergovernmental Panel on Climate Change (IPCC) states, 'the successful implementation of greenhouse gas mitigation options needs to overcome many technical, economic, political, cultural, social, behavioural and/or institutional barriers which prevent the full exploitation of the opportunities of these mitigation options' (IPCC, 2001c). 'In the industrialized countries,' continues the IPCC, 'future opportunities lie primarily in removing social and behavioural barriers' to these technical solutions. But what if merely removing barriers is not enough for realizing these drastic reductions? Perhaps societal 'trend breaks' (substantial shifts in societal developments) are also necessary if we are to achieve a climate-neutral society.

The overall objective of this book is to explore *the need for societal trend breaks to realize long-term substantive reductions in the emissions of greenhouse gases and to assess the prospects far realizing such trend breaks in society.*

This book analyses both the long-term prospects for and short-term implications of the transition towards a climate-neutral society from different perspectives. In so doing, climate change will be placed within the wider context of sustainable development. The basic premise is that given the current situation in climate-change policy and the direction we may have to take in order to realize a climate-neutral society, it will probably also be necessary to change several social trends in a climate-friendly direction — as a precondition for decreasing emissions.

The Netherlands serves as a case study for questions and problems of a similar nature that will arise in other industrialized countries as they are confronted with the challenge of developing long-term domestic climate policies. The main focus of the book is the reduction of greenhouse gas emissions and the possibilities for domestic action within The Netherlands. The impact of climate change on humans and nature is outside the scope of this book, as are possible adaptation strategies.

Some chapters discuss the Dutch situation; other chapters reflect the debate in The Netherlands on strategies that in principle are also relevant for other countries. It is clear that besides domestic action, the so-called flexible or Kyoto Mechanisms (joint implementation, clean development mechanism and emission trading) will play an important role in future reduction strategies. With these instruments, international flexibility is introduced within climate policies, to achieve cost-effective emission reductions. The portfolio of climate policies will always consist of both domestic options and the use of the Kyoto Mechanisms. But given the focus on domestic strategies within The Netherlands, the discussion of the Kyoto Mechanisms will be limited in this book.

Much of the research presented in this book was carried out as part of the Dutch National Research Programme on Global Air Pollution and Climate Change, a strategic research programme in The Netherlands that aims to contribute to the development of long-term climate-change policies. This book is an effort to bring together some of the results of this programme in a coherent framework

The necessity of a climate-neutral society: long-term challenges

Internationally, there is a growing acknowledgement of the need to achieve emission reductions beyond what is agreed upon in the Kyoto Protocol and of the challenges involved in doing so. This awareness arises from a growing body of scientific evidence, which shows the human influence on the climate. The most recent IPCC report (IPCC, 2001a), for instance, states that there is new and stronger evidence that most of the warming observed over the last 50 years is attributable to human activities. The result of this anthropogenic climate change is that society is now forced to reduce its emissions of greenhouse gases into the atmosphere. The United Nations Framework Convention on Climate Change (UN FCCC) states that the world has *'to stabilize the greenhouse gas concentrations in the atmosphere at a level that would prevent dangerous anthropogenic interference with the climate system. Such a level should be achieved within a time frame sufficient to allow ecosystems to adapt naturally to climate change, to ensure that food production is not threatened, and to enable economic development to proceed in a sustainable manner'* (UN FCCC, art. 2).

Ultimately, to achieve a climate-neutral society, the concentrations of greenhouse gases in the atmosphere will have to be stabilized at a new equilibrium, where natural uptake by the oceans and biosphere equals global emissions. This implies that global emissions will have to be cut by half (compared to 1990 emission levels). Historically, industrialized countries have caused most of the observed increases in greenhouse gas concentrations in the atmosphere. Taking equity considerations into account, one could argue that the industrialized world has to make the greatest reductions in future emissions in order to leave 'space' (in greenhouse terms) for development in the South.

Based on several points of departure, such as the maximum allowable temperature change per decade, The Netherlands and the European Union have stated that the concentrations of greenhouse gases should be stabilized before the end of the 21st century at a level well below twice the pre-industrial level. The number of 450 parts per million of volume (ppmv) is often used to quantify a desirable level of CO_2 emissions. The absolute reduction in emissions necessary to realize this level (or any other level) depends on how the world develops. IPCC (2000) devised four future worlds or storylines and, subsequently, six emission scenarios, which reflect different developments in population, economy, energy, technology and land use. These so-called SRES scenarios are a combination of global versus regional orientations and an economic versus an environmental orientation.

The long-term climate challenge for industrialized countries is to transform their carbon-intensive industrial society into a climate-neutral society, and at the same time work with the South in implementing sustainable development there while avoiding large-scale emissions during their economic 'take-off'. The transition towards a climate-neutral society may seem a daunting task, but it is not beyond the bounds of possibility. Chapter 6 of this book presents two visions of how an 80 per cent reduction in emissions (compared to 1990 levels) can be achieved by 2050 in The Netherlands' energy system. The main building blocks of the solutions applied in these two visions are improvements in efficiency, the introduction of sustainable energy sources, reduction of energy demands and CO_2 storage. The two visions are based on the future worlds in the IPCC SRES scenarios A2 and B1 and employ different sets of measures and technologies. They are meant to sketch a spectrum of directions that might lead to the development of a climate-neutral society. In this respect, Chapter 6 provides the reader with a starting point for considering possible technological routes of dealing with the long-term challenges of domestic climate-change policies.

Current climate-change policies

In the context of this book, the current policies on climate change and the short-term dimensions of long-term routes to emission reductions are also important. The current situation is basically the ground on which innovations have to flourish and emission trends have to be reversed if the desired long-term goals are to be realized. It is clear that achieving the short-term climate targets is already proving extremely difficult. In industrialized countries, domestic action has so far not produced very impressive reduction figures. And where trend breaks in emissions have been achieved, this has been the result of unexpected developments (German reunification) or other policy goals, such as the closure of British coal mines, which have not been directly related to climate change. A gap between desires and reality can also be clearly observed in The Netherlands.

Chapter 7 of this book evaluates the current societal support in The Netherlands for climate-change policies and the actual behaviour of various actors, such as businesses, citizens and local policymakers. It explores whether, and if so how, climate-change issues play a role in the decision-making behaviour of societal actors. If, over the coming decades, we are going to realize the long-term levels of stabilization as described above, the current situation will have to change. Chapter 7 concludes with seven dilemmas that actors in society may currently face (to differing degrees) when dealing with climate change.

An important issue in taking action is the variety of perceptions that exist in society about reducing emissions. At one extreme, there is the position that solving the climate problem will have major ramifications and that it will bring industrialized societies to the brink of economic collapse. At the other extreme, the position is that there are many solutions for dealing with the climate problem and that solving the problem is indeed affordable without major economic sacrifices—and there will even be benefits.

In this discussion, it is interesting to note that over the last 20 years, the final energy costs (as a percentage of GNP) for countries that are members of the Organization for Economic Co-operation and Development (OECD) have more than halved and are now in the order of 6–7 per cent of GNP (Blok, 2000). For The Netherlands, it has been calculated that although the costs for reducing CO_2 emissions will increase drastically in absolute terms, the relative costs will only rise slightly (but will stay below 2 per cent GNP) so long as the Dutch GNP continues to rise (Bezinningsgroep Energiebeleid, 2000). IPCC (2001c) comes up with similar numbers for the costs of reaching the Kyoto targets in industrialized countries, but it also points to the fact that lower stabilization levels will lead to higher costs. Since the share of energy in GNP as a whole apparently declines and the economic consequences of climate policies remain limited, one may conclude that this should offer opportunities for developing long-term policies on climate change and for achieving targets. But even this seemingly favourable situation has not led to a reduction of emissions in The Netherlands over the last 10 years.

Achieving trend breaks: some historical examples

In light of the current inability to meet either the agreed-upon short-term targets for reducing emissions or even modest targets, it would appear that the long-term goals for drastic reductions in greenhouse gas emissions are extremely ambitious. If minor reductions are already proving difficult in the short term, will major transitions be achievable in the long run? It may be tempting to think of many relatively easy small steps (annual 1–2 per cent CO_2 reduction) leading to an 80 per cent reduction in 50 years. But the annual growth in the economy must also be considered, since it is directly linked to increasing emissions of greenhouse gases. Radical innovations are therefore necessary—innovations that will require changes in many domains in society.

In implementing the technologies necessary to achieve a climate-neutral society, government and businesses will face many obstacles because technologies will have to compete with established practices and technological trajectories. One should be aware that changes in the order of magnitude discussed here cannot result from central government directives. In the first place, in our modern Western societies, the span of control of national government is diminishing. In the second place, one also has to acknowledge that the type of transitions we discuss in this book are society-wide changes that to a large extent depend on a multitude of diffuse and independent decisions made by individuals, businesses and other societal institutions. Some speak of the *network society* (Castells, 1996: 243–307). It should be noted that these decision-makers have not previously included energy, climate or the environment in their considerations and daily routines and have in many cases not been expected to do so. Achieving significant society-wide changes in greenhouse gas emissions over a period of several decades may imply that 'powerless states' have to take up a stronger role in this network society and its liberalized markets. But governments may no longer be able to exert the influence needed.

Will our modern society be able to deal with the ambitious targets that have been set? It may be tempting to say *no,* but targets that appear too ambitious should not put us off. History provides many examples of radical changes that were made in relatively short periods of time (some even within a few years or a few decades). Environmental policies set goals for drastic changes, including the need for major investments and ongoing innovations to make emission reductions feasible. In 1970, The Netherlands launched a 20-year programme to reduce water pollution, calling for a 60 per cent reduction in pollutants. The Netherlands seems to have a tradition of setting ambitious medium- and long-term targets. In 1989, the first National Environmental Policy Plan in The Netherlands contained a series of targets aimed at reducing a large number of emissions by 70 to 90 per cent within 20 years. More recently, in 1997, the Dutch Minister of the Environment stated that environmental impact per unit of consumption should be reduced by a factor of four in 25 years. Results from the third example can of course not yet be given, but the 60 per cent reduction in water pollution was achieved almost within the allotted time (Vermeulen *et al.*, 1994). The second example refers to 73 industrial emissions to water and air, all with reduction targets in the range of 50–90 per cent for the period 1985–2000. By 1996, 51 of them had already been achieved. It appeared that another II would be achieved in 2000, but the remaining II emissions (including CO_2 turned out to be problematic (Ministerie van VROM and VNO/NCW, 1998).

An impressive example of rapid change can be found in the national energy infrastructure of The Netherlands. This transition demonstrates the need for simultaneous institutional and sociocultural changes alongside technological developments. The Netherlands' energy infrastructure changed from mostly coal-based energy production to gas-based in the mid-1960s. This is an interesting example of a managed transition. It reveals powerful co-operation between business and state, carefully planned over a short period, implemented at the same overwhelming pace and building on the experience and expertise of a few actors in society. Within barely to years of the discovery of large supplies of natural gas in the country, the Dutch energy infrastructure was completely turned around. Yet it is doubtful whether a comparable programme could be implemented today.

The cleaning up of Dutch waters mentioned earlier also involved comparable institutional changes. Reducing water pollution required the development of new technologies, but even more, it required institutional changes within government agencies. The existing structures and institutions were reshaped during this process from small-scale organizations with a single objective, strongly oriented towards agricultural needs, to multi-purpose, more democratic organizations.

A third historical example also illustrates the possibilities and mechanisms of sociocultural change. Smoking behaviour has changed considerably in many Western countries. In The Netherlands, the transition mainly took place between 1970 and 1988, when the number of smokers declined from 60 per cent of the population to 33 per cent, a level that has remained stable since then. In addition to increased taxation, the government used information campaigns stressing the effect of smoking on health. In her study about 'the status of voluntary moderation', Aarts (1998) described this as an example of the 'trickle-down effect' where high-status groups may adopt forms of self-restraint and voluntary moderation as a way of distinguishing themselves from other groups in society. These other groups may then follow the behaviour of the higher-status groups (Aarts, 1998:73, 282). One might expect such mechanisms to be helpful in changing culture, but Aarts is rather pessimistic about this because essential features for far-reaching cultural change are absent (Aarts, 1998).

One lesson to be learned from these examples is that in elaborating the possible paths towards a drastic reduction of greenhouse gas emissions, the establishment of technological scenarios is merely a first step. Trend breaks and transitions are always a co-ordinated development of changes in technology, the economy and society (De Vries, 1994: 296; Jansen, 1994: 499). Most of the chapters in this book address developments in society and focus on structural, institutional and cultural barriers to achieving a climate-neutral society, along with the co-evolution of technology and society to reach this goal.

Using the concept of trends, trend breaks and transition management

In the previous sections, we used the term 'trend break' and we will continue to do so, but it requires some explanation, as does the use of the terms 'transitions' and 'transition management'. This vocabulary springs from Dutch literature on long-term environmental policy-making. It is useful to elucidate some of the discussions on these concepts and the theory of 'trend breaks' and transitions.

In order to realize drastic reductions in CO_2 emissions, societal trend breaks will take place against the background of dominant, almost autonomous, societal developments, such as individualization, demographic trends, internationalization and rapid technological developments (especially in information and communication technology and services). Technological innovations may allow for reduced levels of CO_2 emissions, but one also has to take into account the existence of rebound effects. This is the phenomenon of increased consumption of a service (*e.g.*, lighting or transportation) when the variable cost for that service decreases following an attempt to reduce energy use/emissions through an investment (such as a low-energy lamp or a fuel-efficient heater).

The concept of trend breaks will be used in two ways throughout this book. Trend breaks can be used first of all in a *normative way* to express necessary changes. In this book, the trend breaks required to achieve a climate-neutral society are identified. And second, the concept of trend breaks can be used to *analyse constraints and opportunities* in the transition towards a climate-neutral society. The emphasis in this book will be on this second way of using the concept. We argue with De Vries that it would be rather naive to think that we can 'break' trends if they go against underlying dynamic forces (De Vries, 1994: 295–303). Provoking trend breaks implies identifying such underlying forces and managing simultaneous changes in technology, culture and social and institutional structures to shift the direction of societal developments.

To explore new directions for society (such as the transition to a climate-neutral society) a method called 'back-casting' can be applied. 'Back-casting' can be defined as a participatory method that explores the innovations necessary to realizing a collectively formulated picture of a sustainable future in the long term (Jansen, 1994: 502; Weaver *et al.*, 2000: 121; Jansen, 2000: 173–180). Or, as Robinson (1990) states, 'the major distinguishing characteristic of back-casting analysis is a concern, not with what futures are likely to happen, but how a desirable future can be attained.'

In this approach, back-casting always includes focusing on opportunities and constraints. In this book, images of possible futures (such as those described in Chapter 6) are used in identifying barriers and opportunities for innovation and diffusion, relevant interests, roles and responsibilities, and economic and administrative institutions that will need to be reshaped in order to achieve the long-term target of an 80 per cent reduction in the emission of greenhouse gases.

The next question that arises is how to realize the necessary trend breaks and orchestrate the transition towards a climate-neutral society. In the context of the societal trend breaks necessary for successful development and implementation of the technologies required for a climate-neutral society, it would be wise to open the window of opportunity as wide as possible (Rotmans *et al.*, 2000: 71). In a recent contribution to the debate on long-term policies, Rotmans *et al.* apply the concept of 'transition management', stating that it should be characterized by the following:

- long-term perspectives as a framework for assessing short-term policy;
- reasoning in terms of multiple domains and various actors;
- governance based on learning processes;
- aiming at system innovation and improvements;
- keeping ranges of opportunities open (a broad playing field).

They predict that a drastic transition in the energy system will take at least two generations, based partly on a retrospective analysis of historical transitions from wind to steam (in the 19th century) and from coal to electricity (in the early 20th century). In Chapter 8, possible ways to orchestrate this transition will be analysed and discussed in detail.

The structure of this book

Chapter 6 presents two long-term visions of a climate-neutral energy system in The Netherlands. It looks at the window of opportunity and shows that there are a number of technically feasible ways to realize drastic emission reductions. At the same time, it is well-known that climate change is a tough problem to deal with and that the results of current policies have been relatively meagre. Chapter 7 goes on to identify a number of dilemmas that play a role in current climate-change policies. These dilemmas are elaborated in the light of desired long-term developments. Chapters 6 and 7 together illustrate a sharp contrast between the problematic current situation and the substantive targets that need to be achieved over time. The question, then, is how to realize trend breaks and orchestrate the transition towards a climate-neutral societal. Chapter 8 looks at the governance of technological change and innovation.

Chapters 6, 7, and 8 together form the basis on which the authors of the subsequent chapters have written their analysis about the transition towards a climate-neutral society. As a whole, the book provides the reader with an analysis of prospects of long-term climate policies, short-term issues, the societal trend breaks needed to realize substantive emission reductions and possible ways to realize these trend breaks. However, this book can't provide a complete picture of all relevant elements of such a transition. There are many relevant issues that are beyond the scope of this book, such as the role of business and other new technologies such as genetic engineering, to mention just two.

Chapter 9 analyses households, behaviour and consumption patterns. It assesses future opportunities for drastic changes that would lead to more environmentally sound energy consumption by households and discusses constraints to changes in behaviour. In Chapter 10, the role of local authorities in the transition to a climate-neutral society is discussed.

❑❑❑

6

The Energy System of The Netherlands

Introduction

VARIOUS LONG-TERM PROJECTIONS for the world economy suggest that economic growth is likely to remain strong for many decades. Increased consumption and industrial production will lead to substantial increases in greenhouse gas emissions if current technologies and fossil fuels remain the basis of the energy and production systems. This chapter explores the possibilities of combining steady economic growth with a stringent objective: reducing greenhouse gas emissions in 2050 by 80 per cent compared with 1990 levels. We consider all the main possibilities for reducing greenhouse gas emissions, including increased efficiency in energy and material use, renewables and advanced fossil fuel options, which could play a role within the timeframe considered. This evaluation covers The Netherlands, a country with a high population density and high *per capita* energy use.

In 1998, The Netherlands emitted about 240 Mton CO_2 equivalent. The total emissions can be divided into two groups. The first is made up of non-CO_2 greenhouse gases such as methane (CH_4), N_2O (a greenhouse gas released primarily during the use and production of artificial fertilizers), and CFCs, which together account for about 23 per cent of the warming potential. The remaining 77 per cent are CO_2 emissions, almost exclusively caused by the combustion of fossil fuels to produce electricity and for industrial processes, transport and domestic applications, such as heating. The total energy consumption in 2000 amounted to about

3000 PJ. Over 97 per cent of this energy use is covered by fossil fuels. Renewable energy (solar, wind, biomass, hydro, geothermal) plays a very modest role in the energy supply and covers just 1 per cent (35 PJ, of which 30 PJ from biomass and waste) of the energy demand. The remaining 2 per cent of demand is met by imports of nuclear energy and electricity.

The Netherlands in one of the richer countries within the Organisation for Economic Co-operation and Devlopment, with a GDP of about US$ 365 billion (1999). The energy intensity (energy use per unit of GDP generated) is relatively high compared to other countries. This can be explained by the relatively large share of energy-intensive economic activities, such as heavy industry (for instance, the chemical industry), greenhouse horticulture and transport, in the economy. The Netherlands accounts for roughly 1 per cent of both the world's GDP of approximately US$ 40,000 billion (1999) and the global energy use of about 400 EJ (400,000 PJ).

Furthermore, all these economic activities are concentrated in a small area: The Netherlands covers 3.5 million hectares of land (compared with a global surface area of about 13.2 gigahectares or 13,200 million hectares), or less than 0.03 per cent of the world's land area. Consequently, with its 16 million inhabitants (0.27 per cent of the world's population), The Netherlands is one of the most densely populated countries in the world.

Economic growth has been steady and relatively high over the past decade. Annual growth figures have fluctuated between 2 per cent and over 4 per cent in this period. Although energy efficiency has also gradually improved (as evidenced by lower energy intensities for a wide variety of sectors), the total energy consumption of The Netherlands is rising and there has been no evidence of the decoupling of energy use and economic growth.

Measures to reduce greenhouse gas emissions have not led to a decline in those emissions. Non-CO_2 greenhouse gas emissions are declining, however, mainly because of successful policies targeting industrial emissions and a shift from CFCs to other compounds for various applications. However, this effect is more than offset by rapidly growing CO_2 emissions, mainly due to growing consumption of mineral oil (for transport) and natural gas.

The fact that economic growth has up to now invariably led to higher energy consumption is a worrying starting point, given that economic growth is a worldwide ambition and one that is being achieved. The world economy as a whole has recorded growth rates of 1–3 per cent over the past decade, with much higher figures for countries like China and India. Economic growth is expected to remain high in many developing regions. The world economy could develop and grow at an annual rate of 2–4 per cent. Such figures would lead to an expansion of global GDP by a factor of 4–9 in 2050. Without drastic changes in the way that energy is used and produced, such economic growth will also inevitably lead to a dramatic increase in CO_2 emissions to levels which are without doubt unsustainable for the world's climate.

The IPCC has argued that the risk of human-induced climate change can only be limited to acceptable levels if global greenhouse gas emissions are reduced to around 50 per cent of their 1990 levels (IPCC, 2000) in the course of the 21st century. This is in itself a very ambitious target given the current situation, but it becomes a daunting one if at the same time the world economy develops strongly. The IPCC has also produced a wide variety of scenarios regarding greenhouse gas emissions and the underlying economic and technological developments. While some of those scenarios lead to the desired emission reductions, others give rise to an almost uncontrollable growth of global emissions.

A key principle to emerge from the IPCC exercises is that the industrialized countries have been responsible for the bulk of greenhouse gas emissions in the past and the subsequent build-up of greenhouse gas concentrations in the atmosphere. Developing countries that desperately need economic growth simply to meet basic standards of living should therefore be allowed a larger share of the tolerable emission volume. Consequently a relatively large share of greenhouse gas emission reductions would have to be realized by industrialized countries. Following this rationale, an 80 per cent reduction in greenhouse gas emissions by those countries is often mentioned as a desirable goal for this century.

The key objective of this chapter is to outline a number of quite different ways in which a prosperous industrialized country like The Netherlands could meet this target of reducing its greenhouse gas emissions by 80 per cent by the middle of the 21st century (2050). These scenarios, or 'visions', of the Dutch energy system serve as a point of departure for exploring where in society major transitions are required in order to reach the 80 per cent emission-reduction objective and whether they are feasible or likely. In the course of the discussion in this chapter we will highlight some types of measure and transition that could play an important role. Examples include changes in behaviour and consumption, structural changes in the economy, greater energy and material efficiency, the potential role of renewable energy sources, the use of CO_2 storage when using fossil fuels and measures to reduce greenhouse gases other than CO_2.

The two visions of the Dutch energy system that can result in 80 per cent reduction of greenhouse gas emissions are constructed as follows:

- On the basis of two recent IPCC scenarios (A_2 and B_1) for the world, we have composed two quantitative descriptions of the Dutch economy in 2050. Those descriptions include estimates of parameters at a sectoral level, such as physical production by the major industries, transport kilometers, number of dwellings, etc.
- We have made an inventory and evaluation of options to reduce greenhouse gas that could be available in the long term. This inventory includes options for improving energy efficiency in industry, the built-up environment and transport, a multitude of energy-supply options (renewables, biofuels, nuclear energy, advanced fossil fuel technologies with CO_2 sequestration), as well as options mat lead to reduction of non-CO_2 greenhouse gases.
- Linked to the two scenarios and the principal socio-economic conditions assumed in those two different 'worlds', sets and combinations of measures have been chosen that are likely to play a role in meeting the far-reaching emission-reduction target. For example, renewables play a key role in one vision, hydrogen infrastructure and CO_2 storage a key role in the other.

- Using a quantitative analysis of the selected technology mixes, we have calculated the resulting net greenhouse gas emissions and determined the energy balances for The Netherlands for both visions.

This chapter concludes with a list and summary of topics that are crucial if the radical change in greenhouse gas emissions is to be achieved. Those topics are the key subjects for the subsequent chapters in this book and include aspects such as local and international policy-making, household behaviour and technological innovation. In addition, by presenting two quite different visions for reaching the same objective, this chapter will also provide a basis for identifying major barriers that already exist or can be expected in the future to achieve the radical changes that are required in society. This chapter basically 'sets the agenda' for the authors who will later discuss the various issues in more detail by identifying and discussing the possibilities, barriers, limitations and essential actions relating to the major transitions needed to meet the 80 per cent reduction target.

Two visions of The Netherlands in 2050

It is possible to distinguish several key drivers that determine the level of a society's greenhouse gas emissions. The first major factor is the pace of economic growth, which is directly related to the social and cultural values of a society and the level of consumption and wealth that is considered desirable. A further distinction can be made between a society that gives priority to individual freedom and a more socially-oriented society. Another distinction can be made between a world economy structured on global competition and open markets and an economy with greater government control and oriented towards (regional) trading blocs. Other key variables are the way in which energy and resources are used (in other words, how efficiently they are utilized) and the type of resources (like energy carriers) that are used.

Logically, all those drivers are mutually influential. The predominant value pattern and philosophy is likely to lead to specific preferences for technologies, or perhaps it is better to say that the technologies and resources used are derived indirectly from cultural values.

Table 6.1 Qualitative differences between the visions

	Vision A	*Vision B*
World view	Internationally-oriented "Global Village"	Regionally-oriented, world trade blocks
	Worldwide convergence	
Social environment	Individualistic	Sociable, family-oriented
	Little appreciation of environment and nature	Environmentally-minded
	Personal interest first	Distribution of wealth, social equity
	Well-being of fellow man subordinate to personal interests	Well-being of fellow man important
Economy & consumption	High economic growth and dynamic	Less dynamic and economic growth
	Part of world economy	Part of EU trade bloc
	Motivated by market mechanisms, little government intervention	Regulation, strong government
	Recreation abroad	More nature areas / recreation within The Netherlands
	Quantity above durability	Durable goods, quality
Use of space	Suburbanization	Careful use of space
	Nature areas fragmented	Large continuous nature areas
Traffic and transport	High demand for mobility in passenger and freight transport	High demand for mobility in passenger transport, moderate for freight transport
	A lot of private traffic	Public transport
	Road transport dominant	More freight trains and inland shipping
Energy & environment	Rapid diffusion of technology	Technology develops less quickly
	New infrastructures implemented	Possible solutions on demand-side important
	No change in behaviour	Willingness to adapt behaviour
	Cost-benefit analysis	Environment and sustainability given priority

Table 6.1 gives a summary of the characteristics of two alternative potential futures and different contexts for The Netherlands in 2050. Of course, these combinations of characteristics are in many respects arbitrarily chosen. On the other hand, the factors are combined in such a way as to produce consistent visions.

These 'visions' merely serve as examples to illustrate how a target of 80 per cent reduction of emissions could be achieved in the potentially very different worlds in which it may have to be realized.

The selected visions differ considerably in their assumptions with regard to the situation in The Netherlands in 2050 (behaviour, population, economic and spatial structure, etc.), without actually being 'extreme'. The visions sketch a picture of a conceivable situation in The Netherlands in 2050. The two visions will be referred to simply as 'A' and 'B'.

VISION A

International relations

In this vision, the gap in prosperity between rich and poor countries has disappeared, partly due to rapid developments in transport and communication technology. International co-operation promotes productivity growth and facilitates the faster diffusion of technology. This has led to significant mobility of people and ideas.

The world is characterized by rapid and successful economic development. The most important factor behind the economic dynamism is the trust placed in market-oriented solutions. The level of investment is high, both in technology and in education. The emphasis in this vision is on market mechanisms and increasing productivity, partly thanks to the faster introduction of new and more efficient technology.

Social environment and demographics

People focus on personal progress and want the freedom to control their personal development. Ambition is important: people want to make financial progress and achieve something in their lives. The average standard of education is high. It is important to enjoy life and the level of consumption is high. Concern for nature, the environment and the well-being of others does not have priority. Individualism is the key characteristic of this vision. The workforce includes women as well as men and they work over forty hours a week.

Day care centres are commercial, as is the social service sector. Education is provided mainly by private schools. Steering and policy in society are dominated by technological, market-oriented solutions and approaches. Decision-making processes are fast and there are limited opportunities for public participation.

The family does not have a prominent role in society. The proceeds of the economic growth are used mainly to create further growth, and only to a small extent for social services and the environment and nature. In a world where the differences in prosperity between countries have disappeared, there are still groups within the regions that profit less from the prosperity. Differences in income widen and certain groups fall behind socially. In the multicultural society polarization has therefore taken place. Crime is dealt with repressively.

The world population has grown to nine billion people in 2050. The average age is higher than in 1990. In The Netherlands, the ageing of the population is quite advanced. The Dutch population, at 16.1 million inhabitants, is a little higher than in 1990. The high prosperity results in a long life expectancy and low birth and death rates. In combination with the importance attached to personal development, this has led to a large proportion of single-person and two-person households. The thinning out of the family is evident: The Netherlands has 10 per cent more inhabitants than in 1990 but there are 50 per cent more homes.

Economy and consumption

Due to the dynamics of the free market and the high mobility of people, goods and technology, the world economy (added value) is almost nine times larger in 2050 than in 1990. Although economic growth in the Organisation for Economic Cooperation and Development countries has been slower, total GDP in The

Netherlands is over five times larger than it was in 1990. This implies the same pace of growth as in the last 50 years (including the powerful growth during the post-war reconstruction).

The income per head of the population in The Netherlands in 2050 is five times higher than in 1990. The *per capita* consumption of meat and dairy products is a little higher than in 1990. The most significant changes in spending patterns compared with 1990 are in foreign tourism and the buying of services. People go on foreign holidays several times a year. The aeroplane is by far the most popular form of transport for holidays; there are 15 times as many flights as in 1990. Many day trips and recreational activities are enjoyed within The Netherlands.

The car-ownership rate is high, there is extensive suburbanization and dense transport networks, both national and international. Most passenger and freight transport in The Netherlands takes place by car and truck. The number of kilometers driven for private purposes is 80 per cent higher than in 1990, and freight transport by road is a factor of 8 higher. Freight transport by rail is growing, but passenger transport by rail is lower than in 1990. The volume of inland shipping is roughly one-and-a-half limes what it was in 1990.

Energy and the environment

In this world, possible problems that arise are solved on the basis of a cost-benefit analysis, in which the relevant aspects are assigned a (capital) value. In the process, aspects relating to nature, environment and durability are assigned a low value.

Energy and materials are abundantly available world-wide. This is mainly due to the rapid technical improvements in energy extraction. The energy prices are relatively low and stable so there is little incentive to save energy in this free market. Due to the pace of technological modernization, the prices of solar and wind energy are significantly lower than in 1990 but they still cannot compete with fossil-fuel energy on a large scale. Energy use is high and there is little willingness to pay higher costs for sustainable energy or energy-saving technology.

People are not prepared to adapt their behaviour in order to reduce CO_2 emissions; any measures that have to be taken must have minimal effect on material prosperity and the consumption pattern. People are prepared to pay for measures which allow them to maintain their pattern of consumption. This means that in this world technological solutions are quickly adopted, stimulated by market forces and powerful R&D. Due to the substantial use of fossil fuels, greenhouse gas emissions are mainly reduced by CO_2 storage and the import of (sustainable) energy carriers. Alternatives which fit into this vision are the use of nuclear energy and hydrogen technology. The most important factor behind the adoption of measures to increase material efficiency is the avoidance of the cost of the waste treatment.

Space

People desire comfort and space. Cities are therefore made up of large suburbs where people live in spacious park-like built-up areas. One-third more space is used for living than in 1990, although the population is only 10 per cent larger. Many people in the rapidly growing service sector work at home so that the space required for offices is the same as in 1990. Large industrial complexes take up around 30 per cent more space than in 1990. The Netherlands has several large airports due to the demand for air travel, which is 15 times higher than in 1990. There is a high degree of car ownership and the majority of freight transport is by road. The capacity of the road system is calculated to meet this demand. Broad motorways connect the city suburbs, industrial areas, ports and airports.

Due to the high price of land in The Netherlands the competitive position of the land-bound agriculture has deteriorated. Agricultural acreage has declined by a third in 2050 compared with 1990. Nature areas are exploited to meet the huge demand for recreation. The landscape is characterized by a park-like layout

VISION B

International relations

In Vision B, different regions in the world have developed in different directions. There is less international trade and interaction. There are also fewer powerful international institutions.

In Vision B, the emphasis is on regional identity and self-sufficiency. These important values are developed within the individual cultural and/or economic regions.

Social environment and demographics

Nature is highly valued. The well-being of people and animals and a good environment are regarded as just as important as prosperity. Social aspects play a prominent role in decision-making; people show great concern for others. Attention is given to distribution of wealth and equality in society. The classical family is just as important as in 1990. Men and women participate equally in the labour process and share responsibilities in raising their children. Day care centres and social services are part of the public sector.

The standard of education is high. The government runs education. People are doing well financially and want to contribute to the progress of society as a whole. Personal development is not of primary importance but goes hand in hand with the development of a sustainable and sociable society. People attach importance to quality and sustainability, quantity is less important. Society is structured in such a way that as many people as possible can be involved in decisions. Most of the proceeds of the economic growth are invested in a sustainable economy, social affairs and the environment and nature. Society in The Netherlands is multicultural. Immigrants make up a large part of the working population and are well integrated at all levels of society. Teaching standards and values at local level prevent crime.

In Vision B, the world population in 2050 is 9.4 billion. The population of the Organisation for Economic Co-operation and Development as a whole shows a more moderate growth at 13 per cent than the world average. The Netherlands is an exception: in 2050 the country has 18.9 million inhabitants and has grown some 25 per cent compared to 1990. This is due to both the family-oriented nature of society and the results of an open immigration policy. The Netherlands in 2050 has considerably more elderly people than in 1990. The number of homes in The Netherlands has risen by some 15 per cent. The thinning out of households increases slightly in this vision.

Economy and consumption

In Vision B, the world economy is five times larger in 2050 than in 1990. Despite the greater involvement and the social attitude of the citizens, the relative income differences between the Organisation for Economic Co-operation and Development and the rest of the world still exist. The GDP of The Netherlands is four times that of 1990. The *per capita* income in The Netherlands in 2050 is more than three times higher than in 1990. People go on holiday several times a year. Holiday destinations close to home are particularly popular, but long-distance travel is still in demand. Passenger air transport has increased by a factor of 5 compared with 1990.

The use of cars has risen by 40 per cent per person. Due to the large increase in the population, the total number of kilometers driven in cars is 80 per cent higher. In relation to 1990, the use of public transport has increased by 40 per cent per inhabitant. In total, public transport is used 80 per cent more than in 1990. The bicycle is used for shorter distances. The majority of passenger and freight transport takes place by car and truck, but nine times as much is transported by rail and one and half times as much by inland shipping compared with 1990. Agriculture is considerably less intensive than in 1990. Some 10 per cent less space is used for agricultural purposes in 2050 than in 1990 and, particularly in the cattle-breeding industry, production is lower. Mainly ecological products are consumed and the consumption of meat is lower than in 1990.

Energy and environment

Due to the intensive focus on nature and the environment, environmental problems are seen in the broader context of sustainable development and society. Solutions are first sought at a local or regional level and are aimed at combining the gradual adaptation of behaviour and technology improvements. A great deal of energy is devoted to securing the necessary broad public support.

The world's energy systems differ according to the availability of natural resources in the region. In The Netherlands region, the development of cleaner and more efficient technologies is driven by the notion that we must be economical with energy and material supplies. As a result of the moderate technological advances in energy recovery, the energy prices are relatively high. This is also a motivation for energy saving.

In order to reduce CO_2 emissions, people are prepared to accept measures that can significantly affect their everyday behaviour. People are willing to use public transport and there is a general willingness to pay higher costs for small-scale sustainable energy supplies and ecological products. At the same time, there is rapid diffusion of energy technology with lower emissions (fuel cells, clean energy carriers, combined heat and power), so that local pollution does not occur as readily.

Space

In this vision, people live in busy, compact cities. Within the cities, the concentrated built-up areas are linked up to public transport. Living and working are highly integrated in this vision. The car retains its popularity, but public transport plays an important role in passenger transport. Both the number of motorways and the number of railways have increased since 1990. Air traffic has lost ground to the train for shorter journeys. In this vision, The Netherlands has two national airports to meet the demand for air travel. Due to the large service sector, there are many more offices than in 1990. The space for living and working is used more efficiently but the large population lays claim to the available space. The area devoted to agriculture in 2050 is more than 10 per cent smaller than in 1990. In addition to production, agriculture has an important ecological and recreational function. New features of the landscape are large fields with crops that are used for biomass production (like poplar, willow and miscanthus). Despite the growth of the built-up environment, there are more large nature areas than in 1990.

The visions compared

In vision A of the world, the emphasis is on international co-operation, whereas in B the emphasis is on regional identity and self-sufficiency. People, ideas and capital are less mobile in Vision B than in vision A. There is greater technological innovation in A than in B. In A the world economy is nine times larger than in 1990, compared to a fivefold increase in B. The world's population is somewhat smaller in A than in B, although this change is far more pronounced world-wide than in The Netherlands, which has almost 3 million inhabitants more in Vision B than in Vision A.

Vision A portrays an individualistic, competitive society where people put their own interests above those of others. B, on the other hand, is a sociable and family-oriented vision, where one's fellow man and public interests are central. In the individualistic vision of A, the thinning out of the family will take place far more quickly than in the socially-minded Vision B. The ageing of the population also takes hold quicker in A.

Nature and the environment are used as a springboard for progress in A, and it is mainly for this reason that they are important. In B, nature and the environment come first. The aim is a sustainable society, which does not impose a strain on the environment. Market mechanisms dominate in A, while in B there is strict regulation and distribution of wealth; a consensus dialogue model is favoured for dealing with various socio-economic issues. Vision A, on the other hand, gives few opportunities for participation and has a very hands-off government.

The Dutch economy in A is more than five times larger than in 1990, compared to growth of more than a factor of four in B. However, the population in B is 15 per cent larger than in A. The people in A are thus one and a half times as rich. Due to the larger population size in B, absolute consumption is only a quarter lower than in A.

In B, recreation is largely enjoyed close to home and holidays in The Netherlands are also popular. In A, international travel is part of everyday life throughout the whole of society. In A, people fly two and a half times as often as in B. In B, public transport is available on a much wider scale than in A, and an individual in B travels at least one and a half times more often by public transport. In A, people use their cars some 15 per cent more often than in B, but because there are around 15 per cent more people in Vision B they drive just as many kilometres as in Vision A. Rail transport and inland shipping are growing faster in B than in A. Due to the importance attached to nature, cattle breeding in Vision B is animal-friendly and extensive. In Vision A, agriculture is highly intensive and the well-being of the livestock is not of prime importance.

In B, people are prepared to spend more and to change their behaviour to arrive at a sustainable solution. People in A act as 'homo economics' and are not easily prepared to change their behaviour on the grounds of principle. Measures must have as little effect as possible on their prosperity and consumption patterns.

The people in A live in large houses in spacious suburbs, while in B the people live in compact cities and in smaller houses. There are many motorways in A, while in B there is more of a railway infrastructure. In addition to production, the extensive agriculture in B also has an ecological and recreational function. In B there are large nature areas. The nature areas in A are fragmented by roads, agriculture and built-up areas.

Two energy systems to realize an 80 per cent reduction of greenhouse gas emissions

The energy system of vision A

World trade increases drastically in this vision. The dynamic economic development ensures that there is rapid technological development in many areas. People prefer luxury and comfort and earn more than enough to pay for it. The high incomes of people and the importance of material wealth make it unlikely that lifestyles will be adapted solely in response to environmental arguments. The design of the energy system is largely determined by cost optimization without taking the environmental costs into account. Relatively expensive (energy) options (both for enhancing energy efficiency and energy supply options) are therefore not favoured.

There is a preference for cheap energy carriers. Because of strong competition on the world energy market there are large quantities of cheap fossil fuels available, particularly coal and natural gas (the latter partly from unconventional sources such as Coal Bed Methane and methane stored in clathrates, particularly in deeper sea and ocean waters). Coal, with cost levels at around US$ 1/ GJ, is a preferred fuel, and can be used for the production of both electricity and hydrogen.

The design of the energy system is largely determined by market forces and cost minimization is an important driver. This implies a tendency towards larger-scale production capacity rather than a decentralized energy (electricity) supply system. This production capacity is built close to industrial centres for optimal heat utilization.

Advanced power plants with efficiencies in the range of 50–70 per cent can easily be equipped with CO_2 recovery steps with only moderate economic and efficiency penalties compared to their use without CO_2 capture and storage. Very important is the production of hydrogen from coal to be used for transportation purposes. Carried out on a large scale and combined with CO_2 removal, the hydrogen is carbon-neutral and can be produced with high energy-conversion efficiencies of about 60 per cent.

A significant share of 'coal-based' power generation and hydrogen production comes from Coal Bed Methane. Coal Bed Methane is produced by injecting CO_2 in deep coal layers. CH_4 present in those coal layers is displaced by the CO_2, resulting in a CO_2-neutral gaseous energy carrier.

Nuclear energy also plays a role in the future energy mix, particularly for power generation (30 per cent). Waste heat from nuclear power plants is also used for industrial heat and district heating. Due to the development of inherently safe reactor designs and new methods for treating nuclear wastes (rapid breakdown of long-life isotopes), public resistance to this energy option will have diminished by the year 2050. The low cost of power generation is another argument for the introduction of nuclear reactors. Furthermore, these reactors do not depend solely on uranium for their fuel requirements, but can also be fuelled by thorium.

Electricity probably has to be imported, especially since by the year 2050 a liberalized open energy market has led to a fully functional European grid and free trade in power and gaseous energy carriers. Nuclear-based electricity from France, and possibly also surplus wind power from the UK, can contribute to meeting the Dutch demand.

Another major energy supply option is the import of bio-energy. Produced cheaply in large areas of Eastern Europe, Latin America and Sub-Saharan Africa, both raw biomass and energy carriers from biomass (hydrogen in particular) find their way into the Dutch energy system. Biomass production costs could be in the same range as coal costs. Because there is no need for CO_2 removal and subsequent storage, bio-energy is competitive with fossil fuels. Biomass is used mainly for production of synfuels, and partially for the production of electricity and some feedstock for the chemical industry. Mineral oil remains an important feedstock for the petrochemical industry but not for transportation fuels.

Renewables such as wind and PV play only a modest role. Wind energy is generated mainly offshore in large wind parks but total production makes only a relatively modest contribution of 10 per cent to the electricity supply. The application of PV is limited to niche markets since it generally supplies more expensive electricity than the large centralized capacity. Solar heating makes a very significant contribution by meeting a substantial part of the demand (about 100 PJ) for low-temperature heat in dwellings and offices. The contribution of other renewables is negligible.

The energy system in vision B

There are two very important aspects of the energy system in the B vision. First, the total energy demand in terms of primary energy carriers has not increased due to developments in the economic structure (reduced energy intensity) and far-reaching energy-efficiency measures throughout the economy. The more modest growth in energy consumption and growth of economic activity compared to Vision A puts less strain on the energy infrastructure (distribution of power, heat, gas and fuels). Consequently, the capacity of the energy infrastructure in place at the beginning of the 21st century remains sufficient. This results in a preference for adapting the existing energy infrastructure rather than designing and installing a completely new system.

Secondly, in principle, society prefers renewable energy sources (wind, solar energy and bio-energy). Since renewables (solar and wind in particular) play an important role in the energy supply, and specifically in power generation, the energy system must be able to cope with the intermittent character of those supply options. This involves installing storage capacity and more especially investing in flexible power-generation capacity fired with natural gas (without CO_2 storage, since this would be uneconomical for gas-fired capacity with low load factors).

An advantage of the large share of renewables (linked to a European grid) is that the fluctuations are less severe than for smaller shares linked to a local grid.

The main characteristic of this vision is the strong support for renewables. Nuclear energy is largely missing from the (Dutch) energy-supply mix and the use of fossil fuels is kept to a minimum.

This leads to an impressive increase in the wind-power capacity, both onshore and especially offshore. The longer term potential of wind-power generation is about 15,000 MWe installed capacity. This represents a net contribution to the power generation of about 40 per cent, making wind the single most important source of energy for electricity generation. Another highlight of the power generation in this vision is the structural, large-scale application of solar energy, which is largely concentrated and integrated in the built-up environment (in industry, urban areas and some applications in agriculture). Photovoltaic energy is fully integrated in the built-up environment. PV is still a relatively expensive option for power generation in 2050, but the higher costs are accepted by society. They are partially incorporated in the prices of houses and offices. The contribution of PV to the total power requirement is around 15 per cent.

Solar heating contributes substantially to meeting demand for low-temperature heat (around 150 PJ primary energy).

The bulk of the remaining energy demand is met by bio-energy, which meets two-thirds of the national energy demand. 'Bio-energy refineries', where bio-alcohols are produced as well as hydrocarbons to partially cover the feedstock needs of the chemical industry (the remaining demand is still covered by mineral oil), are the main consumers of biomass. Some of the energy output from these integrated complexes is heat and power. Such complexes are typically built in large, built-up seaport areas where biomass can easily be supplied from all over the world. This reduces supply risks caused by fluctuations in biomass production in different regions. Part of the biomass-fired capacity can also be fed with fossil fuels when needed.

A small part of the biomass (about 15 per cent of the total primary energy demand) originates from indigenous sources, such as (the organic fraction of) waste and agricultural residues. More of it is cultivated, particularly in areas where the growing of biomass can be combined with other land-use functions, such as buffer zones for nature areas and sound insulation barriers. In the case of forestry, the supply of timber, pulpwood and biomass fuel is combined. In total, about 200 PJ can be produced from indigenous biomass.

However, the bulk of the biomass is imported. Imported bio-energy includes raw biomass (for large-scale biomass conversion capacity in seaports) and bio-fuels. Biomass is supplied from various regions around the world, with Eastern Europe, Latin America, Canada and sub-Saharan Africa being the most important suppliers. The Netherlands requires roughly four million hectares abroad to produce the biomass and bio-fuels needed to meet the national energy demand.

Natural gas still plays an important, although relatively modest, role in the energy mix: flexible peak power-generation capacity (highly efficient gas turbines and fuel cells) covers the remaining part of the power production. This capacity makes the large contribution of the intermittent wind and solar-driven power-generation capacity feasible. CO_2 removal and subsequent storage is not considered for this capacity due to the relatively small capacity and low load factors.

Finally, part of the gas supply is covered by Coal Bed Methane, which is produced by injecting CO_2 in deep coal layers. Overall, the use of methane is carbon-neutral since roughly two units of CO_2 are needed to produce one unit of methane. The production potential of Coal Bed Methane in The Netherlands is supposed to be considerable.

In order to operate such a system some CO_2 removal capacity is needed for large-scale energy conversion systems. Part of the CO_2 is recovered at ammonia production plants and other industries where hydrogen is needed and CO_2 is a by product. Some CO_2 is collected from the bio-refineries. As a consequence, CO_2 storage is less important than in Vision A. It is only applied in industrial centres (such as the Rijnmond area) and at caseload power generation capacity.

A comparison of the energy systems in the two visions

This section summarizes some of the main choices, differences and quantitative aspects to be found in the envisioned energy systems of the future. And also summarizes the extent to which greenhouse gas emissions would be reduced with the various main categories of options (efficiency measures, renewables, CO_2 storage) compared to 1990 levels and a baseline development where the growing energy demand arising from economic growth is covered by the present-day fuel mix which is dominated by fossil fuels. Clearly, greenhouse gas emission levels will soar to unacceptable levels if nothing is done to counteract increasing fossil fuel use to drive economic growth. On the other hand, the two very different visions and the strategies that it is assumed will be implemented do meet the target of 80 per cent emission reduction.

Table 6.2 shows that improvements in energy efficiency in all sectors of society are a key factor in achieving this. Even though economic growth is very high over the time-frame considered, in both visions the total primary energy consumption is reduced in relation to 1990. The reduction is particularly extreme in the built-up environment, where energy savings on low-temperature heat for space heating and warm water as well as highly efficient electric appliances reduce the net energy use by around 80 per cent Only in the transport sector does energy use rise in absolute terms compared to the reference year, despite the introduction of much more efficient vehicles and transport systems, as such improvements cannot offset the projected overall growth in transport, particularly freight transport and air traffic.

Table 6.2 Main demand sectors and their share in the total primary energy demand for The Netherlands in 1998 and in the two visions

Demand sector	*Netherlands 1998 (PJ)*	*Vision A (PJ)*	*Vision B (PJ)*
Industry	1203	685	503
Transport	451	874	975
Domestic	538	164	105
Services	482	250	119
Agriculture	192	22	66
Total energy use (approx)	3000	2000	1770

Except for transport, all sectors are able to reduce their total energy demand due to strong energy-efficiency improvements. For transport, energy efficiency improvement is unable to compensate for the (strong) growth in total demand for transport services.

Table 6.3 Simplified energy supply mix for The Netherlands in 1998 and the two visions

Fossil fuels	*Netherlands 1998 (PJ)*	*Vision A (PJ)*	*Vision B (PJ)*
Natural gas	1480	170	130
Mineral oil	1030	220	140
Coal	380	580	150
Subtotal fossil fuels	2890	970	420
Nuclear energy	25	80	-
Renewable energy			
Biomass & organic wastes	30	800	1180
Wind energy	5	30	80
Solar energy	–	110	90
Subtotal renewable	35	940	1350
Total energy use (approx)	3000	2000	1770

Table 6.3 illustrates how the primary energy demands in each of the visions compare with each other and with the 1990 energy consumption. A combination of structural changes in the economy (more emphasis on services and light industrial sectors) and dramatic improvements in energy efficiency make it possible that the total energy demand decreases.

Obviously, the energy mix in both visions has changed dramatically compared to the current situation. The contribution of renewables, in particular (imported) bio-energy, has increased by a factor of 25 (Vision A) or 40 (Vision B). Consequently, the share of fossil fuels has declined, in particular mineral oil. The use of coal (although this could also be true for natural gas) differs greatly between the two visions, although tins fuel is always assumed to be used in combination with CO_2 removal and storage.

Table 6.4 summarizes the key characteristics of the two visions, with an emphasis on the assumptions made and the technologies related to the energy systems that are assumed to be in place. This summary illustrates that the goal of reducing emissions by 80 per cent can be achieved with quite contrasting strategies. Certain aspects appear in both visions (such as increased energy efficiency and mitigation of non-CO_2 greenhouse gases). Other aspects may be competing or even contradictory (such as the energy infrastructure required) and raise the question of how such developments can be steered over time and whether it is possible to determine the optimal strategy in advance.

Table 6.4 Summary of the key characteristics of the assumed energy systems of the two visions

A	B
Energy demand	
Strong growth of industrial production capacity, luxury dwellings in suburban areas and strong growth of individual transport and air traffic are dominant developments affecting the energy demand. With a total population of 16 million people, the total energy demand amounts to 2000 PJ; industry and transport are the dominant sectors in energy use.	More limited growth of industrial production, considerably increased material efficiency, compact cities with highly energy-efficient dwellings, more intensive use of public transport limit the energy use. On the other hand, there are a larger number of people living in The Netherlands, 19 million compared to 16 million in A. Total energy demand amounts to over 1700 PJ; energy use is dominated by the transport sector and to a lesser extent by industry.
Built-up environment	
Large, luxury dwellings in suburban areas. Strong growth of the service sector. Considerable improvements in energy efficiency through good insulation. Energy demand covered principally by electricity and heat pumps and passive solar energy and solar heating systems.	Compact cities and very energy-efficient, compact dwellings. The service sector is dominant sector in the economy. Offices are highly energy-efficient. Energy demand covered principally by heat distribution, solar heating and large-scale application of PV in the built-up environment.
Industry	
Substantial improvements in energy efficiency by new industrial processes. Largely driven by the	Strong improvements in energy efficiency obtained with far-reaching measures, including

(Contd.)

Table 6.4 continued...

A	B
need to reduce production costs and improve product quality. Industrial capacity is organized in integrated industrial complexes with very efficient combined heat and power generation and optimal waste heat utilization. Material efficiency is also improved.	expensive options to reduce energy consumption. Well-developed recycling schemes and adapted products lead to substantially increased material efficiency (up to 40 per cent compared to baseline levels). Industrial production capacity is concentrated in integrated complexes with highly efficient CHP generation and heat cascading.
Transport	
Dramatic increase of air traffic and freight traffic in general. Strong growth of individual transport. Minor role of public transport. Cars and trucks powered by very efficient and clean fuel cell vehicles driven by hydrogen. Efficient aircraft are fuelled by synfuels produced from CO_2-neutral biomass.	Increase in the need for transport in general, but major role for public transport in passenger transport as well as train use for freight. Cars and trucks powered by efficient and very clean fuel cell vehicles fuelled by bio-alcohols.
Agriculture and land use	
Highly intensive agricultural production over a limited surface area. Very efficient greenhouse sector and intensive cattle raising in 'closed stables'. Considerably increased land use for suburban developments (comfortable dwellings in park-like environment). Increased land use for industry and infrastructure (roads, airports). Other land areas important for recreational purposes.	Ecological and more extensive agricultural practice. Integration of functions: agriculture, recreation as well as energy production; land is also used for biomass production for energy and feedstock. Compact cities and large-scale application of public transport.

(Contd.)

Table 6.4 continued...

A	B
Non-CO_2 Greenhouse Gases	
Far-reaching reduction of HFCs through application of alternative compounds, elimination of N_2O emission of combustion-related emissions through application of Fuel Cell Vehicles and other industrial processes. Strong reductions of N_2O and CH_4 emissions in agriculture through closed stables and greenhouses.	Far-reaching reductions of HFCs through application of alternatives, elimination of N_2O emission of combustion-related emissions through application of Fuel Cell Vehicles. Strong reductions of N_2O emissions through less intensive agriculture. Some remaining CH_4 emissions related to cattle raising.
Energy supply	
Energy supply dominated by coal and biomass. Furthermore, a diverse mix including nuclear, natural gas and Coal Bed Methane and wind power. Mineral oil remains of importance for the chemical industry.	Energy supply is dominated by biomass (large-scale import of raw biomass and bio-fuels from other parts of the world), about 15 per cent of the primary energy comes from indigenous biomass and (organic) wastes. Remainder of energy supply covered by wind, solar energy, coal (Coal Bed Methane) and natural gas.
The energy system	
A completely revised energy infrastructure with a large-scale hydrogen distribution network and a network of CO_2 pipelines for collection and storage of CO_2. CO_2 removal is especially applied at hydrogen production and power generation from coal and a number of heavy industries. The energy supply (power, fuels, heat) is dominated by large scale capacity. Energy production is to a large extent integrated in industrial	A more traditional energy infrastructure remains in use: large-scale heat distribution is used for compact cities. CHP generation and heat cascading is widely applied in integrated industrial complexes. The dominant rote of wind and solar energy in the electricity supply requires storage capacity (batteries in the built-up environment, compressed air for wind parks) and flexible, gas fired power generation capacity. PV and

(Contd.)

Table 6.4 continued...

A	B
complexes for optimal utilization of waste heat. The energy supply in the built-up environment is dominated by electricity, application of heat pumps and passive solar energy.	solar heating are widely applied in the built environment.
Electricity	
Electricity production dominated by coal with CO_2 removal and storage, nuclear energy. Remaining part by natural gas + Coal Bed Methane and wind power.	Electricity production for more than half covered by wind and solar (PV) energy. Supportive peak power generation capacity on basis of natural gas and CO_2-neutral Coal Bed Methane.
Fuels	
Large-scale production of hydrogen from coal as well as from biomass. Biomass is also the CO_2-neutial feedstock for production of synfuels, particularly for use in air transport. Substantial import of bio-energy.	Vehicles are fired by bio-alcohols produced from indigenous biomass and large-scale import of biomass and fuels from various world regions. Biomass is also used for production of synfuels, for use in the air transport sector and as feedstock in chemical industry.

Concluding remarks: implications for changes

The exercise above shows a number of things very clearly. In the first place, without measures to limit energy use and promote the use of energy sources with at best very low CO_2 emissions and with steady economic growth over a long period of time, greenhouse gas emissions will soar to unacceptable levels. Secondly, the two fairly contradictory visions show that it is feasible to combine (strong) economic growth with far-reaching reduction of greenhouse gas emissions provided a number of key options are developed and applied. Thirdly, the results also show that such a result can be achieved in different ways. In other words, different, and in a number of respects opposing, strategies are feasible for arriving at a low greenhouse gas emission economy in the long term.

Clearly, the number of (often bold) assumptions made in the visions illustrate that drastic changes in many areas of society are required to reach the greenhouse gas emission levels desired in 2050. However, the changes required may differ considerably depending on which vision one considers most feasible and the context one assumes for the longer term. And changes are extremely complex. Questions immediately arise about how major transitions are to be realized. Different technologies and complete energy systems and infrastructures compete. In addition, technologies become available over different time spans, greatly depending on R&D efforts and (policy) support or international developments. Who will pay the inevitable (additional) costs of transforming the energy system? How are the potential changes to be initiated, implemented and steered?

The visions could be elaborated in far more detail but that is not the objective of this book. The visions serve as illustrative and provocative examples for the later chapters. The visions are the basis for reflecting on the transformations and reversals of (current) trends that are needed to meet the ultimate objective. The next chapter presents an overview that gives some insight into difficulties in realizing crucial changes. They cover social sectors such as households and agriculture, specific options for reducing emissions (illustrated by improving material efficiency as a means to reduce greenhouse gas emissions) and policy areas such as local and international climate policy, technology management and R&D policy and the effectiveness of various policy instruments. The discussion evaluates the feasibility of realizing the drastic changes required to create a climate-friendly society, as well as the prospects of and barriers to success, from very different perspectives. The combination of the findings can provide an insight into transition processes and the full extent of their complexity. An attempt at such a synthesis will be made in the final chapter of this book.

□□□

7

Contemporary Practices in Global Warming

Introduction

THE SCENARIOS SKETCHED in Chapter 6 are just two contrasting examples from a wide variety of conceivable routes towards a climate-neutral society. Whichever route is taken, they all embody manifold drastic transitions within the period of 50 years; technological, institutional, structural, behavioural and cultural.

Such societal transitions will require collective decision-making in the political system, but equally decisions and changes in behaviour by many actors in the sectors of society discussed above. Everyone will have to push in the same direction and that will not be easy to achieve. We know that societal responses vary from denying or ignoring the environmental and societal effects of climate change to adapting or trying to prevent them (Rayner, 1998). Not everyone will necessarily wish to prevent the impact of climate change and to carry the (potentially heavy) burden of a forced transition. Adapting to climate change or even ignoring it may fit very easily into the political culture of substantial groups in society, labelled as fatalists and individualists (Schwarz and Thompson, 1990) or as cornucopians in their attitudes to addressing environmental or climate change issues (see Thompson and Rayner, 1998). The fatalists' and individualists' response to threats of nature like climate change is that nature is capricious or benign. The cornucopians' response is that nature is intrinsically robust and overflowing with abundance of resources. Both responses result in sceptical attitudes.

It may very well be that the political elite of western countries agrees upon the direction and pace of reaching climate goals. However, to get there, they depend heavily on the day-to-day decisions of many different actors, among whom there still is a distinct possibility of meeting scepticism and unfamiliarity with the issue. So the central question at this stage of this book is: what do we know about societal support, about awareness and about contemporary practices in efforts to change the behaviour of individuals and organizations in response to climate change? And if we are confronted with scepticism and unfamiliarity, can we deal with it? Or should we and can we design routes of transition independently of these groups' co-operation? What would that imply for policy scenarios?

In this third chapter we will evaluate the current societal support for climate change policies and the contemporary practices of various actors within society using a large number of studies and evaluations available for The Netherlands. We will examine whether, and if so how, climate change issues play a role in decision-making amongst societal actors in relation to their other interests. We will show that mere is a gap between the willingness to set (modest) targets at the political level (as expressed in various national climate change policy documents) and the slow pace of change in the actual behaviour of target groups (both government agencies and in society). In the last section, we will conclude this chapter by identifying seven dilemmas that are relevant in exploring possible transformation processes and which will be taken into account in the following chapters.

Actors: who decides on innovations?

As Chapter 6 illustrated, a shift towards a climate-neutral society will require the efforts of industries, companies in the energy production and distribution sectors and agriculture. Substantial efforts will also be needed in the fields of transportation and household consumption. Making the transition to a climate-neutral society will not only affect the daily routines and technologies adopted in specific economic sectors, but also represent a challenge for complex processes of collective decision-making, for example in the areas of building and infrastructure projects.

Some of these transitions are large-scale innovations involving few decision-makers. In such cases, businesses depend on co-operation with the national government for approval of projects and perhaps financial support (*e.g.*, CO_2-recovery/storage or large-scale wind parks at sea). Decision-making may then be relatively well-ordered and rest mainly on techno-economic assessments, although local opposition may be encountered and public support might become a critical issue.

But in most cases the transitions depend on far less structured decision-making mechanisms, either involving individual firms or households that do not actually yet address the issues concerned (*e.g.*, applying eco-design principles or using solar energy appliances) and for whom changing tire relevant routines is more than just a matter of making a techno-economic assessment. As Table 7.1 shows, most of the transitions do not involve large-scale projects involving a few actors from business and central government, but fall into the category in the bottom right corner.

This implies that achieving the goal of reducing emissions by 80 per cent mainly depends on the acceptance and willingness of many businesses, local institutions and individuals to contribute to the transitions. This in turn will depend on the perceptions these various groups in society have of their own contribution to the problem and of their own and others' responsibility and possibilities to address it. A complicating factor here is that the suggested options are mutually dependent. For example, the degree to which producers will apply eco-design principles will strongly depend on their perception of the responses of consumers to an increased supply of low impact appliances and consumer goods. As another example, producers' choices in the quality of their energy supply (using renewable energy or not) will have an effect on the environmental profile of their products. This may lead to a growing demand for renewable energy, which will be a strong stimulus for the energy sector to invest in renewable energy. Manufacturers can decide either to leave the choice of modes of renewable energy to energy companies or to be selective at this point and to opt for green energy. In the liberalized energy market, as consumers, businesses may even decide on the modes of energy production, *e.g.*, by producing solar or wind energy themselves. Such

Table 7.1 Scale of the innovative product and number of decision-makers involved in transitions

Scale of	*number of decision-makers*		
product	*few, independent*	*few, mutually dependent*	*many independent*
Large scale	• application of wind (at sea)	• infrastructure projects	
	• CO_2-recovery/ storage		
	• bio-refinery		
	• bio-industry	• application of wind	
	• 50% more eco-efficient aviation	• sustainable technological development policy	• application of biomass • application of CHP
			• hydrogen society
		• fuel cells in motor vehicles	• improving material efficiency
		• widespread reuse/ recycling	• improving energy efficiency
		• low impact construction	• application of sun
small scale		• supply of eco-designed appliances	• eco-agriculture • consumers' choices

interdependencies and growing independencies may form either a barrier to transitions ('it is their responsibility') or a chance to accelerate the process of transition.

Having identified groups of actors that are likely to contribute to trend breaks, we will now roughly sketch the way these target groups are nowadays addressed by policies.

Contemporary policies

The societal sectors mentioned have already been exposed for sometime to environmental and energy policies; policies that have been successful in some fields but not in others. We will show that changes made by these target groups are only partly induced by regulatory pressure. With respect to activities and behaviour related to climate change, the day-to-day decisions of these sectors are often not addressed by compulsory regulations. Contemporary changes in policy and behaviour of these societal sectors are often induced by coercion, voluntary agreements, societal pressure, financial stimuli and market stimuli. In this section we will first discuss the policy strategies applied in the field of energy and environment that address industry, energy production, consumers and mobility. We will then go on to discuss developments in national and local climate change policies.

Environmental and energy policy

Energy policy targeted at *industry* can focus on the energy efficiency of production processes and distribution, possible co-operation in CHP and energy supply and the choice of renewable energy. The actual policy concentrates on improving energy efficiency through voluntary agreements mainly based on energy conservation methods. These agreements are intended to be incorporated in firms' environmental permits. Research has shown that these voluntary agreements are generally effective, although the level of annual efficiency improvement (approx 1.7 per cent in the period 1989–95) will not be sufficient in the context of reducing emissions by 80 per cent (Glasbergen *et al.*, 1997). The most recent generation of agreements is intended to have a wider scope, including wider implementation of potential methods of improving energy efficiency: life cycle management, regional co-operation (industrial ecology), reduction of use of virgin resources and increased use of renewable energy. These methods are also stimulated in *industry-targeted* environmental policy, which also involves voluntary agreements, increased flexibility of permits, application and certification of environmental management systems, stimulation of eco-design, life cycle cooperation and promotion of products with low environmental impacts.

In the field of *energy production and distribution* the institutional changes resulting from the current energy policy with its dual targets (sustainability *and* creation of free energy markets) are clearly visible. The creation of free energy markets for electricity and natural gas requires a substantial re-institutionalization, which is just now taking place. The process involves privatization, resulting in a decline in the influence of local politics. It also implies scale enlargement and internationalization of energy companies, which will possibly affect the environmental policies and practices that these companies developed during the 1990s. On the other hand, new actors are entering the market place, such as retailers and housing agencies selling energy, sometimes solely green energy.

Current policy targets are to increase energy efficiency by 33 per cent in 25 years, the use of renewable energy for 10 per cent of energy demand in 2020 and application of clean technology, combined heat and power (CHP) and CO_2 storage when coal is used for electricity production. In 2001 emissions of CO_2 are still increasing despite the success of the voluntary agreements on energy efficiency in achieving the agreed targets and the reduction in the domestic use of natural gas. Although there has been an increase in the use of renewable energy, it has not been sufficient to meet the policy objectives. Efforts to increase the application of renewable energy have met with a number of problems: local opposition (to wind parks), design and planning processes and procedures in the field of building that do not make provision for it, economic barriers (payback times) and lower prices for non-renewable energy due to liberalization and internationalization of energy markets. In general, the role of government is diminishing, but environmental policy still has an impact through its emission targets and the sector's involvement in the production of policy.

With regard to *consumption,* all consumer goods, transportation and domestic heating are relevant. Policy has already addressed the two latter aspects for sometime. With respect to domestic heating, policy focuses on energy conservation, with the emphasis on financial and communicative strategies. The same financial and communicative strategies are employed to reduce car use. In addition, improvement in the capacity and quality of public transport should further reduce the environmental impacts of mobility. Addressing citizens as conscious consumers employing their 'purchasing power' is a more recent feature of environmental policy which is still in its infancy (Vermeulen, 2000). The second half of the 1990s saw greater use of green taxation, The Netherlands now having one of the highest levels of green taxes. Rising levels of energy taxes are in force for petrol, electricity and natural gas for heating.

Efforts to reduce the impact of *mobility* struggle with the stubbornness of this issue. Reducing the impact may involve the route of addressing motorists' behaviour (applying financial, communicative and physical strategies) or influencing choices of modes of transportation, design of infrastructure, development of clean means of transportation (through support for R&D and regulations). Although these issues are all far from new in the policy arena, on neither of these aspects can substantial improvements, let alone a reversal of trends, be reported up to now, as we are continuously faring increasing levels of mobility. One small positive effect of technological improvement can be mentioned here: the last decade shows stable levels of energy consumption per kilometre by cars, while cars have become larger. There is a rise in the use of public transport, but it barely exceeds the general increase of mobility.

Even more complicated are efforts to achieve more *structural changes,* for instance in the execution of infrastructure projects or large-scale building projects. In these projects decisions are made at the national level and strong businesses are involved. But procedures are long and include many opportunities for reconsideration, both of national policies and local policy implementation, which are likely to be availed of by one or other of the many actors involved. These complex decisions usually require the management of uncertainties relating to economic, societal and ecological effects. This combination of dealing with uncertainty and repeated opportunities for reconsideration can impede the success of such complex processes.

Developments in national climate policies

In addition to general environmental and energy policies, The Netherlands has had a specific climate policy since the late 1980s, We face a conflict between ambition and reality in The Netherlands, where deficiencies in implementation lead to downshifting of ambitions. During the 1970s the Dutch media regularly reported new findings about the consequences of climate change. Between 1979 and 1984 the greenhouse theory was picked up by policy-makers and framed as a CO_2 problem. Between 1984 and 1988 the climate problem rose to environmental priority status, with public attention and political statements of intent both peaking between 1988 and 1990, as was also the case with respect to other environmental problems (Dinkelman, 1995; Dresen en Kwa, 2000; Van der Sluijs, 1997). The Dutch national climate policy started with the international (1988) Toronto target of reducing CO_2 emissions by 30 per cent in 2005.

After 1990 it became increasingly clear that it wouldn't be possible to realize the original targets through domestic implementation. The Dutch National Institute of Public Health and the Environment (RIVM) forecasts that with current policies the emissions of greenhouse gases will continue to grow. With the economic growth The Netherlands experienced over the last few years and current policies, The Netherlands can only meet its target under the Kyoto Agreement if an extra package of measures is implemented. Furthermore, reductions will have to be bought abroad (Milieuverkenning 5, 2000–30).

The picture that emerges from this overview is, on the one hand, one of good intentions and a willingness for The Netherlands to be at the forefront of (international) climate policy-making. But on the other hand, it shows a country in a difficult position to realize its climate targets. The latter point can be explained by the economic structure of The Netherlands, with its energy-intensive industry and agriculture, the importance of transport (see also Chapter 6) and its enormous dependence on what happens internationally. In light of this situation and the awareness of the long-term necessity for far-reaching measures, in 1996 the Dutch parliament started (the first) parliamentary hearings on climate change. Parliament wanted to be in a better position to form its own opinion about the climate problem and the measures that were needed. The debate about the very existence of the climate problem and the uncertainties surrounding its causes and effects played a major role in these hearings. These hearings resulted in a political re-affirmation of the seriousness of the problem and the statement that the committee of enquiry considered a reduction of greenhouse gas emissions by 30–40 per cent in 2020 (compared to 1990 levels) to be possible.

Interestingly enough, research by Ester *et al.* (1999) highlights a remarkable difference between what decision-makers think the Dutch public wants and what the public say themselves about their preferences. Policy-makers often complain about what they see as a gap between their habit of looking at strategic, global issues and the more locally-oriented environmental issues that occupy the minds of the general public. The argument is that citizens are only interested in environmental issues insofar as they affect their day-to-day lives. The research by Ester *et al.* shows that this perception is not entirely accurate. An example: all decision-makers say they are well-informed about climate change issues but they believe that only a minority (35 per cent) of the public would say they are well-informed.

In fact 80 per cent say they are well-informed (whether this is correct is not important here). Even more interesting is the perceived willingness to provide extra resources for environmental protection. On this question, given a choice between seven general policy issues, 39 per cent of the policy-makers put the environment in first or second place. They thought that only 7 per cent of the public would do so (and that 0 per cent would place it first). In fact, public support is far larger, as 27 per cent gave environmental protection a first or second place (Ester, 1999).

This history shows that climate change is a widely recognized environmental issue in the national policy arena in The Netherlands. Furthermore, decision-makers seem to underestimate the societal support for environmental policies. But at the same time, far-reaching targets for 'a climate-neutral society' are absent from official policy documents. Such a long-term approach to the climate problem has not been widely discussed in the political debate in The Netherlands up to now, although more recently some influential advisory bodies have been demanding more attention for this aspect (VROM Raad, 1998; AER, 1999; SER, 1999). At the end of 1998 the Dutch Environmental Advisory Council published an advice entitled 'Transition towards a low-carbon energy system' (VROM Raad, 1998). In this advice it called for structural steps in the long-term transition towards emission reductions of 80 per cent in industrialized countries. It argued that this long-term target should drive the policy process, in which structural steps towards these goals are fundamentally more important than short-term successes.

In this context, the process in the COOL project is another interesting sign of growing attention to long-term climate targets among stakeholders in The Netherlands.

Local climate policies

A similar conflict between ambition and actual practice can be observed at the local level. Local authorities can play an important role in realizing national climate policy targets, but can also make more ambitious policies. As in many other countries, municipalities in The Netherlands have responsibilities and decision-making powers which are essential for achieving the transitions discussed in Chapter 6. For instance, they have to issue permits for any new construction or form of land use. They are responsible for environmental permits for smaller firms. They have an important position in local traffic and public transport policies. But in all these fields their ability to pursue climate policy goals is limited because powers and responsibilities are in many cases shared with provincial and national authorities. In many cases, budgets have to be provided by national authorities. Climate change has received attention in many municipalities in recent years. Some have incorporated it in Local Agenda 21 projects, where municipalities have discussed environmental policies with local stakeholders. Other municipalities have committed themselves to the International Climate Alliance of 1999, in which 109 Dutch municipalities (which is 20 per cent of all municipalities in The Netherlands) agreed to develop policies aimed at 50 per cent reduction of greenhouse gases in 2010 compared to 1987 levels. Members co-operate in developing possible policies and exchange information.

Although climate change discussions in Local Agenda 21 and participation in the Climate Alliance can be seen as positive signs, we must not overestimate their impact. An evaluation of the local climate policies of 51 municipalities in 1999 showed that in many cases consideration of climate change in strategic policy formulation is no guarantee of changes in greenhouse gas-related activities by local actors. There are small groups of communities actively engaged in implementing climate change strategies, but research shows that in most cases possible actions in the field of construction, traffic and green energy are not taken in practice. In many instances, basic information on sources of greenhouse gases is lacking, developments are not monitored and plans are not even developed (Burger, 1999). Frequently, a few attractive examples of activities, such as subsidizing solar hot water systems, are highlighted but there is rarely a comprehensive strategy. Research into planning procedures for large building locations and urban reconstruction projects reveals that, despite existing local policy plans, environmental and energy policies are not integrated in development procedures from the start of such projects. Many of the existing 'demonstration examples' were promoted at a later stage by actors (like energy agencies, energy companies or consultancies) who were initially not included in the process of plan development (Van der Waals *et al.*, 1999; Van der Waals *et al.*, 2000).

The main barrier at this stage of development of local climate change policies is (what is called) external integration of climate change issues in other fields of local policy, like traffic, economic development, urban and land use planning, housing, tax policy, etc. On the local level we see a considerable degree of political support for climate policies, but up to now this has only partially resulted in changes in local sectoral policies and physical changes. Opportunities for local climate policies will be discussed in more detail in Chapter 10.

Contemporary practices

There is some research into the attitudes to and public support for climate policy and contemporary practices of various groups in society. However, the picture is not complete and many different approaches have been adopted. Still, we believe that the overall picture shows a significant difference between declared public support for climate policy and daily practices and routines. We will discuss the available information in the context of public opinion, producers and consumers.

Public opinion

In the late 1990s it was often stated in public debate in The Netherlands that public concern with environmental issues was declining because of the rising prosperity. Empirical evidence doesn't support this statement when it is expressed this simplistically. Annual environmental behaviour monitoring shows that in the second half of the 1990s the Dutch population increasingly saw environmental problems as an important issue. Between 1995 and 1999 the percentage that stated that environment is 'one of the most important issues' rose from 40 per cent to 47 per cent. The number of people saying it is 'the most important issue' ranged between 17 and 20 per cent. Equally remarkable was the (small) growth in the number of people who said that environmental issues would be the greatest threat in the future: 27 per cent in 1995 and 33 per cent in 1999 (Couvert and Reuling, 2000).

What we are actually seeing is a strong shift in the perception of environmental issues. They are decreasingly seen as threat to the health of this and future generations and increasingly people feel that 'the fuss about the environment is exaggerated' (Couvert and Reuling, 2000). So public opinion is tending to shift away from an apocalyptic interpretation. We are also seeing a shift in opinions on responsibility. Dutch environmental policy is built on the principle of shared responsibility of producers, retailers and consumers. Public opinion, however, increasingly holds industry responsible and individuals are less inclined to see themselves as being responsible (Couvert and Reuling, 2000).

This perception of personal responsibility may be the most important explanation for the gap between attitude and actual behaviour. Analyses show that this gap is smallest with respect to waste treatment and energy use and water use. The gap is widest in the spheres of food, mobility and purchasing behaviour (Hoefnagel *et al.*, 1996; Becker *et al.*, 1996; Steg, 1999; RIVM, 1998 a and b; Couvert and Reuling, 1999), so actual consumer behaviour is still unaffected.

At this point we should be aware of a fairly strong societal differentiation along the lines of income and even more of education. Research into environmental attitudes and behaviour shows a dear differentiation, especially according to educational levels (Hoefnagel, 1996).

The discussion above is not specific to climate change issues. There is little research specifically on the issue. One example is an analysis of public knowledge about causes, effects and solutions for environmental issues, which showed that in 1998 73 per cent were able to give a cause of climate change correctly, 65 per cent could mention a correct effect, but only 31 per cent could give a correct solution. However, looking at the full chain of causes, effects and solutions, only 25 per cent could give a proper combination, whereas 18 per cent couldn't give a proper presentation of any part of the chain (Couvert and Reuling, 1999). This analysis covered a period of four years. It showed no significant increase in the level of knowledge during that period (Couvert and Reuling, 2000). Yet other research shows that the public itself believes that of all environmental issues it is best informed about climate change, with 80 per cent saying they are well-informed or very well-informed (Ester *et al.*, 1999).

With specific reference to energy conservation, research shows that the majority feel this is an important issue (56 per cent), with only 1 per cent saying it is 'not' or 'not at all' important. The number of people who feel it to be important is declining. On this aspect we also see the societal differentiation along the lines of income, and even more the level of education: people with a lower income and level of education explain energy conservation behaviour with economic arguments (squandering of energy costs money), whereas people with a higher income and level of education tend to stress environmental arguments (Couvert and Reuling, 2000). Another study analysed priorities given to various environmental issues and showed that climate change is given low priority in public opinion: 8 per cent and 11 per cent giving it first and second place respectively. Overpopulation (28 and 12 per cent) and air pollution (38 and 27 per cent) were given high priority by a far larger proportion of the Dutch population! Accordingly, climate change is not broadly perceived as being dangerous for human health (Ester *et al.*, 1999).

Why is information about public opinion important in this context? Because it is an essential resource for the necessary political process and is crucial for changing consumption patterns. Here we can see two alternative routes. The first is the route of 'conscious consumption', in which citizens may affect producers' willingness to improve the eco-efficiency of their products by purchasing alternative products with substantially lower environmental impacts. Following this route, the need for consumption is not addressed. In the second route the aim is to reduce the volume of material consumption by advocating less materialist forms of consumption or retrenching.

The latter route has been explored by Aarts, who studied rare examples of a culture of retrenching in material consumption and followed other sociological examples of refraining from consumption (like smoking). One might also expect the trickle-down effects identified by Elias to occur in the field of environmentally-relevant behaviour. Material consumption trickles down from groups with a higher social status and better education to the rest of society (Aarts, 1999). Her argument is that such a trickle-down effect is not likely to occur with respect to a culture of retrenchment because it lacks essential features, like the ability to distinguish yourself from others through this type of behaviour (lack of prestige) and because such an environmental attitude does not represent a cultural capacity contributing to obtaining power (Aarts, 1999).

This discussion raises the issue of affecting lifestyles. Aarts cites Breemhaar *et al.* (1995), who showed that consumers assess their alternatives in the light of their sets of final values (like freedom, autonomy, health, personal development, social appreciation, etc.). Environmental issues are seldom included in these sets of final values (Breemhaar *et al.*, 1995). This line of argument opens the way for research into social mechanisms for constructing sets of final values in combination with social differentiation of lifestyles (see *e.g.,* Schwarz and Thompson, 1990; Motivation, 1999), an approach scarcely followed in social environmental research up to now.

At this point, we conclude that there is public support for addressing climate change, at least at the level of attitudes. However, the public doesn't respond to the greenhouse issue as such but to environmental issues in general. Up to now, the climate issue plays scarcely any role in day-to-day behaviour and there are clear signals that it isn't given a high priority. There is little to be expected from pleas for sober lifestyles.

Producers

In the 1990s impressive improvements occurred in the general environmental performance of producers. In the late 1980s Dutch environmental policy set ambitious targets for the period 1990–2010; in some respects demanding an 80 per cent reduction in an even shorter period than that discussed in this book for CO_2. Producers were addressed with a mixture of three policy strategies: a central approach using modern forms of coercion and incentives, an interactive approach and internalization, and a self-regulation approach (Vermeulen, 2000a). In 1998 the Dutch government and producers together published a book entitled *The Silent Revolution* singing the praises of their 'co-operative efforts in improving the environment'. In Chapter 5 we said that these targets will be met (Ministerie van VROM/VNO-NCW, 1998b; RIVM, 1999). This success is supported by the institutional changes in businesses during the 1990s. We are witnessing a continuing diffusion of institutional change. The number of companies that have integrated environmental management into the business process or are in the process of doing so has grown from 37 per cent in 1997 to 49 per cent in 1999 (KPMG Milieu/NIPO, 1999: 24; see also KPMC Milieu/IVA, 1996). More recently, we have seen rapid growth in the number of firms with ISO 14001-certified environmental management systems (ISO, 2000).

We should be aware that, by implementing environmental management systems, companies in theory put in place the institutional requirements to meet the governmental and societal demands in all areas of environmental improvement, including climate change. However, during the 1990s climate change and energy-efficiency policies have followed different paths, being instigated by different government departments and employing different policy instruments. During the 1990s the Minister of Economic Affairs, who is responsible for energy policy, established a separate forum for consultation with the aim of reaching voluntary agreements on improvements in energy efficiency. As we discussed in the first section of this chapter, this approach has proved effective. Despite low energy prices, companies have met the agreed targets. Research shows that they are more motivated and have improved their organizational capacity to increase their energy efficiency, and energy efficiency has taken a more prominent place in their decision-making procedures (Glasbergen *et al.*, 1997; Glasbergen, 1998).

The first generation of energy-efficiency agreements was mainly concerned with optimization of production processes, application of combined heat and power and energy conservation. They did not cover the use of renewable energy. Natural gas and electricity are simply taken from the national grid. Large industrial firms deserve special attention as large-user tariffs are an interesting issue to address, especially since the European energy market is in the process of liberalization. In 2000 only a handful of businesses used renewable energy, like photovoltaic cells, wind energy or electricity with green certificates (examples are the national rail company NS, Van Melle, a manufacturer of sweets, and Siemens, as producer of wind turbines). In doing so they have faced opposition. For instance, Van Melle could not get local permits to place wind turbines at any of its sites in The Netherlands.

Recent research on 'cleaner production' shows us what types of measure are being applied by businesses. Depending on the type of measure, between 25 per cent and 65 per cent of companies have applied specific measures. Most of these involve a reduction in the use of materials and adaptations in production processes, but there have also been energy-related measures (between 60 and 65 per cent of the companies) (KPMG Milieu/NIPO, 1999).

There is little specific information about businesses' attitudes towards the climate change issue and climate change abatement. A study from 1995 involving over 500 business decision-makers in The Netherlands (Research and Marketing, 1995) shows that about one in four were strongly committed to their environmental performance and one in five showed a strong commitment to climate change issues. However, other (more traditional) environmental issues are regarded as far more important. In general, the level of knowledge and perceptions of climate change issues differ little from those of the general public (see also the previous section on public opinion). Only one in three expected effects on society in the short term. Their main source of information was mass media, but two in five said they generally ignored articles in newspapers or programmes on television about this issue.

Nevertheless, business decision-makers generally see industry as one of the main perpetrators (85 per cent) and it is generally accepted that industry has a leading role to play in abating climate change (89 per cent), while also expressing great confidence in the (technological) capacity to resolve the problem. The main barriers to solving the climate change problem are felt to be economic interests, conflicting policy interests and the inadequate pace of technological development. About two-thirds of those surveyed stated that their own company should take initiatives. Regrettably, the study didn't look in more detail into the willingness of companies to act themselves and what they did in practice. A more recent study (Van der Woerd *et al.*, 2000) reported a tendency amongst multinationals in the US and Europe to move towards a more proactive approach towards the climate problem.

Another study focuses more specifically on investment behaviour by firms (Gillissen *et al.*, 1995) and describes barriers to diffusion, the investment criteria that are applied and policy preferences. Their analysis shows that 20–30 per cent of the technological potential for energy conservation is unknown to industry, even in the case of profitable technologies. Perceived impacts on production quality and continuity (possibly due to 'start-up' problems of new technologies) are important barriers and it was found that firms' perceptions of profitability differ significantly from profitability calculated in policy studies (Gillissen *et al.*, 1995).

Our observation here can be that there are visible changes in modes of production and in the integration of environmental management approaches. There is some difference between large (internationally-oriented) firms and SMEs. Multinationals like Shell and BP-Amoco are currently shifting their position and are leaving traditional lobbying organizations against climate policies and moving towards a more proactive position. Energy-efficiency is receiving growing attention, but is often not yet integrated in environmental management. Even when it is, it has to compete with other cost-reducing (or liability-reducing) environmental issues. Scarcely any attention is paid to 'renewable energy'. Chapter 8 will discuss possible routes for the acceleration of dematerialization and technological development.

Consumers

Consumer behaviour is relevant for CO_2 emissions. Energy use in households for heating, electricity and use of appliances has been the subject of energy policy for nearly a quarter of a century now. The growing mobility of consumers adds substantially to the volume of CO_2 emissions they generate. And, more recently, attention has also shifted to the entire volume of consumed goods. Every product bought may use energy during domestic use, but also includes energy used for its production and disposal. This indirect energy requirement of products and services consumed actually exceeds the energy requirement for domestic use of electricity, heating and use of petrol for cars (Biesiot and Moll, 1995).

Governments have been attempting to influence domestic use of electricity and heating and the use of petrol for cars since the mid-1970s, when The Netherlands was boycotted by Arab oil producers. Many lessons have been learned, but it would take too long to discuss them all here (see Piers Consultancy, 1992). The best way to see whether policies were effective is to look at physical effects and market information. These show that the use of electricity rose during the 1990s at an average of 3.4 per cent per year. The use of natural gas is declining due to the successful promotion of insulation and high-efficiency central-heating appliances (Brezet, 1994).

In the field of mobility we are far from a trend break: the number of kilometres driven has been growing at 2 per cent per year, although the rate of increase seems to be slowing. That is explained by a dramatic growth of congestion on the roads. We see a gradual growth in the use of public transport (3.5 per cent per year), but it still only represents a small proportion of the volume of private transport (about 1:7 to 8) (RIVM, 1998b; CBS Statline).

The other aspect being considered here, the entire volume of consumed goods, is worth looking at in more detail. Biesiot and Moll studied the energy requirements of consumption patterns. This study shows a strong relation between income levels and total energy requirements: households with an income of €10,000 on average use twice as much energy as households with an income of €5,000. But even more interestingly, we see large differences within all income categories: those who use a lot of energy use twice as much as the thrifty users within the same income group (Biesiot and Moll, 1995). Although this may be largely explained by physical factors (quality of the houses) and situational factors (distance to work), choices in consumption present quite significant opportunities for greenhouse gas reduction by means of conscious use of products and services with a low energy requirement. Another study shows that motorists themselves believe that 20 per cent of their car use is avoidable (Tertoolen, 1994). A large-scale experiment in The Netherlands showed that with intensive efforts to raise awareness, consumers could combine increased wealth with a reduction of total energy requirements by 30–40 per cent (Schmidt and Postma, 1999).

Results in real-life situations give far less reason for hope. Tertoolen tried to influence car users' behaviour through persuasion with arguments concerning the collective and personal disadvantages on the one hand, and by improving self-monitoring, feedback and commitment on the other. He concluded that car users will not stop voluntarily because car use is predominantly linked with feelings of independence, ease and comfort (Tertoolen, 1994).

When we look at purchasing behaviour, we can use information from a longitudinal study of environment-related behaviour (Couvert and Reuling, 2000). It showed that 10 per cent of the population often takes the environment into account in purchasing decisions and 20 per cent do so sometimes. The others just choose on functional or economic grounds. In the period 1993–99, these levels were generally stable for most types of product in the study. There was a minor tendency among high-income groups and environmentally-oriented individuals to improve their 'performance.' In the last year, low-income groups and people who are indifferent towards the environment tend to drop out with respect to purchasing products with less effect on environment and separation of waste.

Here again it is most instructive to look at information available about the effects at the end of the production and consumption chain, or to put it another way the end of the *information decision behaviour* chain. The study discussed above is based on what people say they do. The best proof of behaviour is to look at what is actually being sold. Looking at purchasing behaviour, we see only very small market shares for eco-labelled products. As regards food (representing half of the total indirect energy requirement of households (Biesiot and Moll, 1995), the market share of milk produced under the European EKO label for organic food products grew from 0.6 per cent in 1998 to 1.5 per cent in 1999 after supermarkets changed their policies. Comparable figures for other food products show us three things (Vermeulen, 2000):

1. far more people say in surveys that they buy organic food than actually do;
2. actual market shares for organic food are very small. But there is a small glimmer of hope;
3. changes in policies can have large effects in market niches.

The central question here is whether addressing consumers with both information about the relative environmental impact of alternative products and improving the supply of the environmentally improved goods will bring about trend breaks. The non-governmental organization Global Action Plan employs a method of information transfer, self-monitoring and social stimulation in assisting small groups of citizens with changing their behaviour. These groups are called eco-teams. Drawing on evaluations of eco-team behaviour, they claim to be able to realize a 25 per cent reduction of CO_2 emissions (GAP, 1999). Others like Aarts and Breemhaar, whose arguments we discussed in the section on public opinion, are far less optimistic.

These observations have recently led to what might be called a withdrawal from addressing citizens as conscious consumers. The Dutch National Advisory Council for the Environment recently stated in its advisory opinion on 'sustainable consumption' that it is unwise to expect consumers to buy more expensive product alternatives with better environmental performance. Urging them to do so would imply an overly moralizing role for government. Instead, emphasis should be placed on addressing producers (Raad voor het Milieubeheer, 1996).

Our main conclusions from this section are that we are seeing a shift from an approach where consumers are addressed mainly with respect to energy conservation through their behaviour (heating, lighting, buying low-energy household applications) to one that addresses consumption as a whole. Recent environmental policy has addressed citizens as consumers, but with marginal results. Promoting sober lifestyles tends to be ineffective as it is too contrary to dominant values. Willingness to make voluntary financial sacrifices is small. Rising abatement costs might collide with low acceptance within segments of the population (related to level of education and lifestyles). At the same time, government's attention for the consumers' role may need to be reinforced.

Reflection: main dilemmas

Surveying the research discussed above, we can see a widespread recognition and acceptance of climate change as a problem that has to be dealt with. This applies to both the level of policy-making and the level of society (stakeholders and public opinion) in The Netherlands. The 1990s also highlighted a willingness to set quantitative policy targets, and at the local level even quite substantial ones. However, they address the short term and involve a step-by-step approach. Because of the difficulties experienced in changing energy-related behaviour, targets have been moderated and even those are hard to achieve.

In addition, looking at the actual behaviour of the relevant societal actors, we see a gap between a willingness to act at the political level and the slow change in actual behaviour of target groups or, in other respects, the absence of the climate change as an issue in their daily lives.

We have consciously focused on the situation in the 1990s to show the contrast between the long-term images shown in the previous chapter and the current situation in Dutch society. Of course this only gives us a picture of attitudes and actual behaviour for a specific and short period, and shows a difficult struggle with very modest policy targets. Over the longer term, perceptions, attitudes and daily routines may prove to be less irreversible than suggested by these observations. However, in our quest for societal barriers in the development towards arriving at the images presented in Chapter 6, we are better off taking into account these constraints and finding ways of dealing with them. Even if we assume that the climate change problem is recognized, resistance among different groups in society based on considerations of cost-effectiveness and competing interests and goals will have to be faced. On the other hand, in our open, non-directive society, long-term and substantial climate change policy will inevitably require co-operation by these different groups in society. Knowing this, we can identify various obstacles to securing public willingness to act:

- a substantial segment of public opinion doesn't support major interventions (relation with income levels, lifestyles and specific political cultures);
- re allocation of administrative competences implies that national governments may have political targets but will increasingly depend on others in attaining these goals;
- in local and regional politics substantive action towards greenhouse gas reduction has to compete with more urgent and inescapable local priorities. This in a context where effects of climate change are barely perceptible and local authorities will rarely be the key actors in combating climate change;
- for many (but not all) of the societal groups involved, other 'basic' values than environmental consciousness construct the initial motives behind choices and orientations of daily behaviour. Conduct relating to greenhouse gases will have to be pushed into these 'basic values-driven' behavioural routines or will have to produce co-benefits;
- there also appear to be some obstacles within industry. There are clear tendencies towards improvement of environmental performance, but for this group of stakeholders greenhouse issues have to compete with other environmental and socio-ethical issues, often involving more direct responsibilities.

The following chapters will explore the most relevant routes for social transformation and identify necessary trend breaks in social institutions, cultures and practices. The authors will adopt different academic perspectives and focus on various sectors of society. The obstacles discussed above confront us with a number of difficult choices one might not want to make between alternative approaches: seven dilemmas in defining strategies for achieving a climate-neutral society. These dilemmas are presented next.

1. *The dilemma of a 'small number of decision-makers' trajectory* versus *a 'decentralized, multiple and diffuse decisions' trajectory*

Do we prefer to realize an 80 per cent reduction by means of a 'central-technological' trajectory, choosing to apply large-scale technological solutions and clean production, or with a trajectory involving as many societal actors as possible, all asked to do their bit to solve the problem in their roles as consumer, motorist, producer, investor, decision-maker, etc.? The second route may be very sensible, opening up as much creativity in society as possible, but it requires sufficient involvement of society. On the other hand, while the 'small number of decision-makers' trajectory may very well be feasible (*e.g.*, to build enormous wind parks in the North Sea), it may reinforce the weak public support and may even contribute to the growing distance between politics and people.

2. *The dilemma of 'acting now'* versus *'delayed response'*

   This dilemma refers to the timing of climate policies. Long-term targets will determine the necessary efforts. According to some, if an 80 per cent emission reduction is to be achieved we have to act immediately. The priority needs to be on reducing emissions and stimulating technological developments in the coming 15–20 years. 'Optimists', on the other hand, believe that a delayed response is appropriate. Future technological progress will enable rapid, cheap and drastic emission reductions in the future, so there is no reason to intensify climate policies right away.

3. *The dilemma of 'focusing on climate change'* versus *'pursuing sustainability'*

   The public does not perceive climate change as a separate problem. This non-exclusivity of the climate change problem applies equally to many of the technological solutions and policy options. Pushing through a – 80 per cent transition may require separate and selective attention in the form of special campaigns and policies on climate change, especially when target groups (still) scarcely acknowledge the problem or the sacrifices needed. Separate public campaigns, separate routes to solutions, clash with existing preferences for attacking climate change in integrated environmental, or preferably sustainable development, approaches.

4. *The dilemma of 'demystification'* versus *'need for increased awareness of change impacts'*

   Research into environmental attitudes shows a gradual demystification of apocalyptic messages concerning the environment. In the context of 80 per cent reduction of greenhouse gases, the lack of support for climate change issues at the level of genuine action calls for campaigns designed to increase understanding of climate change and the need to abate it. In that case, apocalyptic messages tend to be counter-productive.

5. *The dilemma of 'central steering via direct regulation and market incentives'* versus *'co-production'*

   Regulation and market incentives in theory seem to be effective if enforcement is well organized and incentives are substantial enough. However, due to the administrative burden and real locative effects, both forms of central steering have serious difficulties in emerging from the political decision-making process unscathed. Participatory approaches in developing climate change policies (co-production) may be the solution, but they require clear politically determined targets and a context of direct regulation and market incentives, discouraging free riders.

6. *The dilemma of 'transition through co-benefits'* versus *'inevitability of a special climate change approach'*

   Opting for solutions that are attractive thanks to the co-benefits (simultaneous benefits of solutions, like cost reduction, functional qualities, comfort, safety, etc.) makes it possible to ignore lack of societal support for climate change policies alone. However, although this may help in the first stages on the way to an 80 per cent reduction, it will probably not be sufficient in itself. A special climate change approach may be inevitable.

7. *The dilemma of 'decentralization and Europeanization'* versus *'the need for national direction of the transition'*

Achieving an 80 per cent reduction will not only require clear choices in technological fields and modes of enforcing their implementation, but also constant monitoring, activation and revision of policies. We tend to assign this role to national governments, but due to two bi-directional processes, decentralization and Europeanization, European national governments have fewer tools in their own hands and depend more and more on the co-operation of others.

Too many dilemmas may make it hard to see whether we will ever be able to make the transitions described in the previous chapter. However, the practices, policy initiatives, policies under construction, partial support in the various groups in society discussed above could equally be seen as a breeding ground for initiatives that, given fertile growing conditions, will flourish in due course. The following chapters will explore the necessary *societal* trend breaks, consciously bearing in mind the tensions between ambitions and recalcitrant societal practices.

❑❑❑

8

Technological Change and Innovation for Climate Protection

Introduction

THE CHALLENGE OF CLIMATE CHANGE demands radical changes in technology and new directions in innovation to satisfy the technological needs of the climate-neutral economy. According to the IPCC analysis, the Western economies may have to reduce greenhouse emissions by 80 per cent in 50 years, a very short period on the timescale of technological development. The future world with 80 per cent fewer CO_2 emissions sketched in the A and B visions in Chapter 6 of this book clearly indicates the technologies needed in such a world. The future world calls for technologies that differ in almost every respect from the technologies currently used in all sectors of the economy. The question facing modern industrialized society is, therefore, how to initiate such technological changes, whether they should be incremental or radical, and what the role and contribution of the actors involved, in particular government, is or should be. Although governments are not the only actors charged with seeing through the change process, in industrialized society they bear a special responsibility for guiding future developments. Only democratic governments possess democratically-legitimated power to make authorized decisions in this regard, because of the normative nature and content of such decisions. But governments cannot and should not work alone in the change process towards a climate-neutral economy.

Our analyses will show that desired climate-neutral technologies cannot be ordered or chosen as large or small scale technological options (compare dilemma 1). Radical change emerges from constant and continued search for better technologies. Furthermore, technological development is by definition a multi-actor process and driven by the many goals and ambitions of the actors involved and not by the desire to reduce the climate change problem. What can and should be done, therefore, is to work on the internalization of climate protection needs in these multi-actor processes. In this way climate protection might become a new value orientation of the many, many actors carrying innovation and a new guiding principle in the search for the next technological steps. Each new step might incubate a radical change and, therefore, it makes no sense to wait and hope for radical change in the future (compare dilemma 2). Internalization of climate change needs in processes of technological innovation and change is a long-term process by definition and requires diversified governmental actions and a rich policy repertoire both at the national and the international level (compare dilemma 5). Though we might know the kind of climate-neutral technologies needed, the process to initiate and to develop these kind of desired technologies cannot be managed top down. The process itself is by definition highly differentiated, quite often of a global scale with many, many different actors involved, each influencing the course and direction the process takes in terms of new technological steps. Policies, therefore, are constantly facing the dilemma to focus explicitly on climate change needs or to tackle these needs more indirectly as co-benefit (compare dilemma 6). What is best cannot be prescribed, but should be decided by a clear analysis of the technical and non-technical inertia involved in processes of socio-technical change. In the context of this chapter it is only possible to highlight some of the general aspects of socio-technical change processes from an evolutionary point of view.

The evolutionary view of technical change is based on the idea of innovation as part of systems and trajectories. With the help of this idea it is possible:

1. to analyse how and why technology predominantly changes along fairly incremental paths of development and why it is so difficult to initiate more radical types

of innovation. In the next section incremental change in technology will be explored in more detail.

2. to explain what types of innovation are needed with a view to reducing CO_2 emissions by 80 per cent in 2050. The A and B visions assume fairly radical innovation of existing technologies, such as a 40 per cent increase in material efficiency in industrial production and consumption. Therefore, next we discuss the implications of these kinds of improvements in terms of system innovation and transitional change.
3. to identify new points of intervention and types of policy. In particular, the evolutionary view stresses the need for policy-makers to be actively involved in the variation-selection processes that change technology, *e.g.*, by stimulating the invention of new technology and the selection of climate-friendly technology in the economic process. Based on this, we go into the implications of an evolutionary-based analysis of technological change for technology policy. Our analysis will show a need to design and implement short-term policy strategies from a longer-term transitional perspective. Finally, the last section of this chapter summarizes the major conclusions of the analysis.

Lessons from history: why technology tends to change incrementally

One of the key lessons taught by scholars of technology dynamic is the high degree of complexity and the collective nature of technological change processes, which makes them fairly difficult to influence, for instance from a climate change perspective (although not impossible, as will be shown in later sections of this chapter). The scientific understanding of this complexity is still rather fragmented, but literature refers, among other things, to the multi-layered nature of technological change and to the solid embedding of technology in society as factors contributing to complexity. Technologies are part of functional systems (food production, energy distribution, etc.) that have evolved, and their evolution has given rise to beliefs, institutions, infrastructures and interests that guide the actions and the thinking of the actors. So technical change does not occur in leaps. A new technological invention initially tends to develop step-by-step overtime with small improvements of its technical and economic performance. This is because technological development builds cumulatively on past achievements, ideas and knowledge. Past achievements are not only visible in the technologies and technology systems around us today, but also in scientific laws, technological principles, science and engineering practices and search routines through the training and education of new generations of scientists and engineers. Technology builds on this cumulative knowledge. Individual innovations also depend on complementary technologies For their production and use. In this way, technologies and technology systems build and develop their own irreversible patterns of development which focus and narrow the way we search for new and better-performing technologies (Rosenberg, 1982). In technical change, as in human life, there are path-dependencies ('lock-in').

If we repackage the empirical materials in more theoretical concepts, we might say that technological changes take place within a material context and a cognitive and normative framework. Each societal function (such as production, transport and mobility, heating and illumination) is satisfied within a specific socio-technical system. For our transport and mobility needs we developed car-based and truck-based transportation systems, for our production needs we developed specific industrial systems (iron and steel, agriculture, chemistry) and for our power needs we built electricity systems. The systems developed in specific ways and according to specific patterns, while technological change within these systems was guided by the cognitive and normative framework, which is known as a 'technological regime' (Nelson and Winter, 1977; Kemp *et al.*, 1998). The regime consists of beliefs and rules that guide the actions of actors.

Actors invest in technological change according to 'rules' or 'heuristics'. These consist of the engineering consensus about the relevant problems and approaches to solving them, assumptions about what users want, ideas about how the system should be organized, and the roles of actors. Actual decisions regarding technological change that are taken at the micro-level are strongly influenced by the rules of a regime and not just the rules of the market. If a regime is stable, technological change can become relatively autonomous in the sense of predictable; actors look for technological change in a familiar direction (*e.g.*, improving car technology rather than changing car-based transport itself).

The path-dependent nature of technological change is not just a matter of engineering consensus about the relevant problems and approaches to solving them, but also depends on the value of investments in particular technology systems and the costs of building new systems. The examples of technological change within mature manufacturing industries such as pulp and paper and iron and steel illustrate this. The existing layout of capital assets is used as the starting point for technological change within these 'sector' regimes. The main production route is to optimize the existing system to deliver products at the lowest possible cost. This example illustrates how and why existing and dominant regimes tend to induce 'incremental' innovation in the majority of socio-technical systems. Such innovations lead to 'system optimization' rather than 'system innovation' (see also the section on long term technological change below). System innovations occur if society manages to satisfy certain functional needs in a completely new way, i.e. in ways that strongly deviate from the prevailing socio-technical system. In many cases, these types of fundamental change occur through an accumulation of several related innovations; together inducing the emergence of a new technological regime. In this way, existing regimes are changed, often with quite a lot of reluctance, for a new set of principles, rules and practices, which will guide further technological change in the system.

Innovation and technological development take time, almost always years or decades, as the example of the shoe-press technology in the paper industry illustrates. This is because technological change and development is not a programmed, or top-down process. Change processes involve millions of actors and their daily decisions and activities. Neither climate change problems nor other societal needs and demands guide these decisions. At the micro level, decisions are guided by the wide variety of economic, institutional, scientific and engineering rules that form a technological regime. Actors are motivated and guided by the economic logic of the market, the political logic of the bureaucracy or the scientific logic of the laboratory. A variety of incentives, past experiences and future expectations motivate and influence these actors in their decisions and activities that jointly shape technological change. Because of the many actors involved and the rule-guided nature of technical change, technology develops in a step-by-step (incremental) and patterned fashion. Technology tends to change in specific directions, along specific paths of development, which tend to endogenously proceed in certain directions. Changing or redirecting these courses and paths, for instance in climate-friendly directions, proves rather difficult, but not impossible, as will be shown in the following sections of this chapter.

Modulating dynamics: long-term technological change

Incremental movement along the route of technological development will probably not bring about the economy with 80 per cent fewer CO_2 emissions depicted in the A and B visions. Both visions assume the emergence of radical system innovation. System innovation involves changes in technology systems beyond a change in parts or components of technology. It is associated with new linkages, new knowledge, different rules and roles, a new 'logic of appropriateness', and sometimes new organizations. Historical examples of system innovation include the electrification of manufacturing and homes, the development of the centralized electricity system, the cluster of synthetic materials innovations and machinery for injection moulding and extrusion. A system innovation of our own time is e-commerce, with the Internet as a new infrastructure for communication and Internet providers as new agents. It should be noted that system innovation does not necessarily imply change in every component or at every level of the 'former' system. System innovation usually consists of a combination of new and old components and may even consist of a novel combination of old components.

The notion of system innovation is highly relevant for the topic of this book (mitigating the climate change problem) because system innovation is one way of achieving environmental benefits and other types of sustainability benefits (in terms of natural capital being preserved, health gains and improved social well-being).

The Dutch DTO programme has identified several sustainable technologies (technology systems) that offer efficiency improvements of considerable magnitude. These include novel protein foods (improvement by a factor of 10–30), precision agriculture (improvement by up to a factor of 50), decentralized production of electricity using renewables and microturbines, underground transport of commodities in pipelines (improvement by a factor of 10 in energy efficiency) and industrial ecology. All these solutions offer benefits for climate protection in the form of a substantial reduction of greenhouse gas emissions, although they are not, apart from renewables, viewed as climate-protection technologies. Their development could be supported from a climate-protection perspective, which requires the coordination of climate protection policies with sector policies, an issue we will discuss at the end of this chapter.

The distinction between system innovation and system optimization is useful because it forces us to think about the long-term consequences of innovations: whether or not they give rise to or contribute to system innovation or alter the current path of development.

Personalized public transport, such as dial-a-ride services, and CO_2 policies are believed to contribute to system innovation in the form of intermodal travel in which people combine individual means of transport with collective means. Anti-congestion policies are believed to sustain the current trajectory of motorized passenger transport based on the individual use of cars. Some innovations may be part of both system optimization and system innovation (renewal). An example is urban cars, which may be used in combination with collective forms of transport for long trips or as a second or third household car. It is not uncommon that innovations can be used both within an existing system and within a new system. Very often system change (what we call system innovation) emerges out of existing systems. A good example is the development of the computer regime, which has evolved from the regime when computers were used solely for computing activities and have since found new types of application—for process control, word processing and information communication (through the Internet).

Quite often radical innovations are first used in a 'conservative way' within an existing system before they become an element of the new system. Gas turbines are an example of this. They emerged in the electricity system to meet system needs, but at the same time they opened up new technological options for decentralized power generation outside the central station system (see previous section on incremental technological change).

Transitional changes are thus the outcome of innovations and processes that can be labelled as system optimization or system innovation, although which category the innovation falls into may not be clear from the outset and may change with time. Such transitions occur very slowly. They usually take one or two generations (25 to 50 years) and they are not deterministic technological change processes; that is, they cannot be predicted with any kind of precision. But the change processes that make up the transition can be seen as occurring within one of four different transition phases.

1. a *pre-development phase* of dynamic equilibrium where the *status quo* does not visibly change;
2. a *take-off phase* where the process of change gets underway because the state of the system begins to shift;
3. an *acceleration phase* where visible structural changes take place through an accumulation of sociocultural, economic, ecological and institutional changes, all of which influence each other. In the acceleration phase there are processes of collective learning, diffusion and embedding;
4. a *stabilization phase* where the pace of social change slows and a new dynamic equilibrium is reached.

The distinction between different types of innovation is relevant for (the management of) environmental technology responses. Environmental protection benefits may be achieved along a route of system optimization or innovation. The first route facilitates quick results, and is usually favoured by the sector and policy-makers. The latter route of system innovation offers the prospect of long-term benefits through long-term structural change.

Between the two, we have a partial system redesign as an intermediate option or an end option. Some people have made a case for incremental change. For example, Clayton *et al.* (1999) and Berkhout and Smith (1998) have argued that significant environmental benefits may be achieved through incremental change over an extended period. Whilst this is undoubtedly true, at a certain point in time such changes will run into increasing marginal costs per unit of improvement If we want to achieve 10- to 50-fold improvements in resource productivity – which according to some are needed in the next 50 years – we need system innovation or technological regime shifts (involving structural change) in addition to the optimization of existing systems or product chains (Ayres and Simonis, 1994; Kemp and Soete, 1992; Rotmans *et al.*, 2000).

The A and B visions of a low-carbon economy described in this book reflect a range of innovations which are not labelled in terms of 'system optimization' and 'system innovation' but in terms of strategic visions. The strategic visions are made up of innovations and policy measures and changes in the cultural and technological landscape, all of which constitute the building blocks of a new socio-technical system emitting 80 per cent less greenhouse gas in 2050. But it is not really clear how to use these visions for policy and management of the change process. What do they imply for technology policy or any other type of policy? Should we have technology-specific policies or generic policies that change the framework for economic decision-making and technical change? And what does it imply for the government-industry relationship? What is government's role: that of a facilitator, process manager, director or mediator? The next section will turn to these governance questions in more detail.

The governance challenge

The patterns and dynamics of technological change sketched in the previous sections not only indicated the complexities involved but also showed that government is just one of the actors trying to influence the pace and direction of technological change. Change processes are fuelled by millions of decisions of actors and government cannot decisively influence technological change, *let alone* control and manage such change processes top-down. Initiating the changes needed to bring about the future climate-neutral economy sketched in the A and B visions is therefore basically a challenge of *governance* of socio-technical change processes rather than just a challenge for *government's* technology policy. Governance refers to decisions and actions of both public and private actors in processes of technological change, whereas government refers to democratically legitimized collective decision-making and policy action. In confronting the climate change challenge, it is of great importance to develop governmental strategies capable of supporting, strengthening and redirecting the governance of the many actors influencing socio-technical change processes in climate-friendly directions.

Two major conclusions from our brief sketch of technology dynamics above could be a starting point for developing these kinds of strategies. These conclusions are:

1. reliance on the autonomy of actors and factors influencing socio-technical change and innovation is not sufficient for a drastic reduction of CO_2 emissions in the next 50 years. This is likely to lead to predominantly incremental changes (system optimization);
2. new technologies and new technological paths (system innovation) do emerge, but the transitional potential of this kind of innovation is hard to identify at the time of their emergence.

   Based on these conclusions it is possible to set out some general lines for developing technology-oriented policy strategies.

- *First*, technology policy should continue to stimulate short-term (incremental) technological innovation, but preferably those innovations incorporating potential for system innovation.
- *Second*, two types of reference systems should be used to identify the potential for system innovation of incremental innovations: a short-term reference system and a longer-term transitional (system innovation) reference system.

- *Third*, short-term technology policies should take account not only of the contingencies influencing technological change in the short term but also refer to and fit in with a longer-term transitional perspective of system innovation.

This section sketches some implications for redirecting technological change processes in more climate-friendly directions.

A contingency view of policy

The major lesson that technology studies have to offer for short-term technology policy is that the effect of government intervention depends on the type of technology being considered and the regime in which technological change takes place. Stimulating the development and introduction of photovoltaic energy requires a different government strategy from stimulating the development of energy-efficient technologies in mature industrial sectors. The most important point of departure for developing technology policies is that innovative technologies differ, economic sectors differ, existing technological systems and regimes ruling those systems differ, and hence processes of technological change towards a more sustainable mode of operation differ. There is no standard recipe for directing or accelerating technological change in more climate-friendly directions.

If a government wants to enhance climate-friendly technological developments, it should be clear about the various regimes of technological systems, the actors they can approach and the dynamics of technological change (what drives technological change?). This is the only way to evaluate whether governmental strategies will have an effect, and if so how they will have the greatest effect. Some innovations have a strong enough momentum of their own (for example, strip-casting technology for the steel industry or shoe-press technology for the paper industry) and scarcely need any governmental intervention. Drawing on the particularities of actors, the technologies and the technological regimes as well as the economic sectors should be taken as a point of departure for developing arid implementing technology policies. Policies should take into account the whole set of contingencies fuelling technological change.

The choice between technology-specific and generic policies

System innovation requires special support. In appraising innovations eligible for support, one has to ask: does the innovation that produces benefits for climate protection improve the existing regime or offer possibilities for inducing a 'system innovation'? Is there a need for support, and if so how should it be organized? In defining policies, the following elements should be considered:

1. the improvement in environmental performance of an innovation, *e.g.*, strict CO_2 reduction potential;
2. the contribution of the innovation to system innovation. In what respect does the innovation deviate from existing regimes?

In answering these questions one should consider four aspects:

- to what extent does an innovation physically fit into an existing technological system? Does it replace or complement the existing technological system? Can it be used in both systems?
- does the innovation have a large or small effect on other dominant rules / performance characteristics in the regime? What is the dominant rule in the regime? Is this rule undermined?
- does the innovation offer possibilities for further improvement, for achieving learning economies? Does it have negative effects? Are there ways to deal with them? Is there a need for the innovation or not?
- does the innovation only affect specific firms or entire product chains and/or industrial sectors?

3. are there developments in the economic, social and technological landscape that sustain the innovation? What are these developments and what are their prospects?
4. who is actually involved in developing the innovation? Why? Who will benefit from the innovation? Is there a need for support?
5. which actors can government approach (whom can they influence?) to actually engage in technological change?

What government intervention strategy best fits the existing network of actors involved?

Such an assessment of the incremental or radical nature of an innovation's potential can only be tentative. It needs clear understanding of the existing regime as a point of reference. Such an analysis leads to insight into the extent of the changes that are needed to get an innovation introduced; such an analysis goes beyond a static analysis of the costs and performance of a specific innovation. It needs a comprehensive analysis of the regime and of the feasibility of altering regimes. The first type of analysis consists of mapping the roles, outlook and basic assumptions of regime actors; the second type of analysis involves the identification of alternative futures and visions that are attractive from the perspective of climate protection and related societal points of view. Actors in the existing regime may have long-term visions that are interesting from a climate-protection perspective. Within the car industry, for instance, there is a converging view that in the future people will be using vehicles that run on alternative fuels, and that car use will probably no longer imply individual car ownership but instead car rental. Bill Ford, chairman of Ford, the world's second-largest car maker, has said that fuel cells will end the 100-year reign of the internal combustion engine and that we may witness an end to car ownership as the preferred method of personal transportation. In this vision, Ford and other car makers could own vehicles and make them available to fee-paying motorists when they need access to transportation (*Wall Street Journal Europe,* 6th October, 2000). Government policies for climate protection could build on such ideas, because the emergence of this kind of system innovation will need governmental support. In the example given, the innovation relates not only to new technology but also to structural changes in the car-based mobility system. One of the major challenges would be to develop a car-rental system that meets the same standards of flexibility and quality as private ownership of cars.

The actual design of government policies for climate protection is no simple matter. Policies may be disruptive, or even unnecessary. An example might illustrate this. If government wants to stimulate energy-efficiency improvements in the paper industry it should take into account the concentration of R&D capacity in the relevant machine-supply industry. There are only two major European companies that supply machinery for the paper industry. The two companies almost totally control the core of the exiting production technology for paper production. These industries can therefore contribute to energy-efficiency improvements from the 'inside out.' One should also bear in mind that because of their reputation and credibility, these two well-established actors are best positioned to push the traditionally conservative paper manufacturers towards innovation. Governments that want to encourage technological change in the core of the components used in paper manufacturing have to address these suppliers. These actors should not necessarily be stimulated with R&D support since a number of innovative technologies currently being developed have other (more important) advantages than energy saving. The risk of free-rider behaviour (getting money for doing things they would have done anyway) is considerable. Furthermore, it is even questionable whether they are interested in government support to fund the core of their R&D activities. An alternative might be a sort of voluntary agreement on technology development at European level. In this way, governments can try to establish energy efficiency as a higher priority on these firms' research agendas. This is just an example of how government can fine-tune intervention strategies in the existing regime. A more detailed look at the current regime of paper production also makes it clear that the slow rate of technology component replacement at a paper plant is ultimately a major obstacle to exploiting the energy-efficiency potential of 'innovative' technologies.

For these and other kinds of intervention strategies there is already a rich variety of policy instruments. Table 8.1 presents a possible classification of government intervention strategies and instruments.

For the purpose of this book, the question is of course whether the strategies and instruments presented in Table 8.1 can initiate the transitional changes needed to achieve a low-CO_2 economy. The section on long term technological change showed that technological transitions are associated with structural change at different levels—companies, production chains, consumers and governmental policies—and are linked to new ideas, beliefs and sometimes even new norms and values. Many of the elements

Table 8.1 Policy approaches and strategies

	Supporting *Technology variation*	Actor Interaction + learning *Coupling of variation and selection*	Forcing *Market selection*
generic	– Subsidies for R&D	– Eduction policy – Innovation centres	– Co_2 taxes – tradeable permits, – covenants about energy saving
specific	– Research programmes for climate-friendly innovation – Subsidies for adoption of climate-friendly technologies – Government procurement of green innovation	– Societal discussions about sustainable transport, energy, etc. – Development of strategic visions and technology road maps – Strategic niche management	– Goals for renewable energy – Fuel economy standards

involved in transitions, such as lifestyle changes, rising income levels and changes in beliefs and values, cannot be managed. This raises the question: can a transition towards a low-CO_2 economy really be managed? The simple answer to this question is of course 'no, transitions defy control'; a more qualified answer is that although transitions cannot strictly speaking be managed, it is possible to work towards a transition, to give leverage to processes of change, and to *modulate* ongoing dynamics.

Basically there are two ways to accomplish this. Firstly, through generic policies, for example by Judiciously applying economic and/or social incentives and disincentives to make some possible paths more interesting and feasible than others and *vice versa*. Secondly, through technology-specific policies, for example the promotion of electric vehicles through research programmes, tax credits (investment subsidies) and the use of local experiments. Technology-specific policies interfere directly in the dynamics of technical change, rather than indirectly as generic policies do. Examples of generic and specific policies are given in Table 8.1. They are arranged into three categories that are derived from an evolutionary perspective of processes of socio-technical change: policies that stimulate technology variation, those that work through the selection process and those that couple variation and selection.

For a transition to a climate-friendly economy, both types of policy approaches are needed. Generic policies such as carbon taxes are needed, especially for the uptake of climate-friendly technologies. They are in line with the economic dictum that 'prices should speak the environmental truth'. Products should be penalized according to their economic cost to society. Prices should reflect social costs. Positive externalities should be rewarded. This is an argument for the use of subsidies for technologies whose social benefits exceed the benefits for the innovator, as is often the case with climate-friendly technologies whose environmental benefits have no value in the market place.

Generic policies that change the economic frame conditions for technical change are important for transition processes towards a low-carbon economy. However, it is questionable whether they alone are able to promote the development of radical solutions that have not benefited from dynamic and scale and learning effects and adaptation of society (Kemp and Soete, 1992; Kemp, 1994). Radical solutions are often rejected and opposed by regime actors who favour incremental change (system optimization) because it is less disruptive. Therefore, the promotion of radical solutions and system innovation also requires technology-specific policies. So both types of policies are needed, and the pros and cons of each of them will differ from case to case.

Even then, we do not feel the combined approach of generic and specific policies is enough to produce the transitional changes required for the climate-neutral economy of 2050. System inertia can be strong, as our analysis in the previous sections of this chapter showed, and fossil fuels are still considered extremely significant for industrialized society and its development. What is lacking are clear notions of alternatives for the fossil-based development paths in industrialized society and clear perspectives on alternative technological routes. These kinds of future climate-neutral perspectives could provide a meaningful context for the development and implementation of short-term technology policies, either generic or specific. In combination with short-term policies they can fuel transitional change. What is needed, therefore, is a comprehensive new perspective on climate-oriented technology policies to align short-term policies to longer-term goals. Transition management provides such a comprehensive, integrative perspective.

The need to align short-term policies to a longer-term perspective: transition management

Drawing on the analysis of transitions in technology dynamics and systems analysis, scholars have suggested transition management as a new approach to interactively processing transitional changes (Rotmans, Kemp *et al.*, 2000). Transition management offers a methodology for organizing and processing transitional change while avoiding a deterministic and narrowly goal-oriented focus on the future. Transition management has been developed as an alternative for this kind of deterministic and goal-driven policy approach, an approach that in fact ignores uncertainties by setting policy goals based on or derived from deterministic, and often unrealistic, blueprints for future scenarios. The majority of these blueprints are unable to account for the complexities involved in transitional change. A basic premise of transition management, therefore, is that the processing of a transition should depart from and account for the uncertainties regarding the direction and content of future developments, as well as the complexities in terms of interdependencies, causalities and relationships between ecological, social and technical systems. It has been suggested that transition management accounts for these complexities coherently, understands them and develops robust and sustainable solutions.

Transition management, therefore, actively fosters transitional change by means of interaction between governmental and private actors and is process-and content-oriented. The process is a device for interactively learning how a transition might appear and for learning new practices from the perspective of the transition (learning by doing and doing by learning).

The notion of transition management offers a framework for theory and action. Transition management is seen not as a task for government but as a task for society as a whole, whose problem-solving capabilities, imagination and support must be mobilized. At the same time, it is believed that for transitions offering collective benefits rather than private benefits for the individuals concerned, as in the case of a low-emission energy system, government has an important role to play in initiating, fostering and sustaining a transition. This role is twofold: to realize certain substantive objectives such as CO_2 reduction (this is the *content* role) and to make sure that the process of variation-selection works well (this may be called the *process* role). The latter is important for securing desirable outcomes in the long term. The process role is aimed at stimulating and organizing the transition process, mobilizing the social actors concerned, creating opportunities and challenges for participants in the transition and creating boundary conditions within which the transition process can operate.

The transition goal is by definition multi-dimensional, and should take into account the different positions and ambitions of actors participating in the transition process. Furthermore, the transition goal should allow for more than one trajectory of development since one of the important aims of the process is to learn about future development options and not to achieve one particular long-term goal. Transition objectives therefore consist of a *basket of objectives* and is informed by quality images articulated by the actors taking part in the transition process.

The transitional track can be designed using the so-called corridor approach, not in terms of a fixed blueprint but in terms of acceptable societal risks. Societal risks can be formulated in terms of economic, sociocultural and environmental risks (boundaries) and these risks provide a kind of dynamic corridor specifying the boundaries for future developments. These boundaries are themselves not fixed but are constantly adjusted in the context of the progress with the transition process. In this way, the corridor provides a dynamic set of frame conditions for shorter-term policy development and policies in terms of interim goals. These interim goals should not only be formulated in terms of material content, for instance, x per cent CO_2 reduction, but should also refer to achievements in the transition process itself, for instance in terms of learning effects. The interim goals should be evaluated and adjusted during so-called development rounds, when actors participating in the process analyse, assess and evaluate the learning effects regarding both process and content in the longer-term transitional perspective. Based on these evaluations, adjustments can be made regarding the transitional process, material content and the policies needed. In this way, transition management guarantees societal support, active participation by societal actors in the process and in development rounds, and the use of their knowledge and experience in the transitional process.

As already mentioned, government assumes a special position and role in the process, a role that could be labelled 'acting director.'. This role combines aspects such as organizing the process, stimulating and mobilizing actor participation, process facilitation and the like. This role as director must be combined with an acting role, referring to the government's democratic responsibility for developing and implementing policies for transitional change. The different phases in the transitional process could determine the policies needed for both aspects of the governmental role as actor and director. In the pre-development phase of transitional change the governance challenge is to create a broad playing field and not just to rely on the options that are available. It is important to explore radical options. Here the goal of learning is more important than the actual greenhouse gas emission reductions achieved. Suggestions for how to do this are described in the literature on strategic niche management (Kemp *et al.*, 1998; Weber *et al.*, 1998; Weber and Dorda, 1999). In the take-off phase, the governance challenge concentrates on actor mobilization and motivation, for instance the interactive development of transitional goals and images. In the acceleration phase, the governance challenge is to facilitate the development and diffusion of innovations, both technical and non-technical (organizational, institutional, economic), to facilitate societal embedding of new developments and to support learning. In the stabilization phase, the governance challenge is to facilitate and support consolidation.

Learning and societal embedding are important elements of socio-technical change and hence an important aim of transition management. A thorough analysis of actors is a condition for the success of a transition process, and needs to be complemented by studies of trends and ongoing developments in niches and regimes. The various action perspectives (outlooks) of the actors should be traced and made explicit. Not only the perspectives of the existing, influential actors, but also those of the less important players who might become important in the future should be considered, as actions undertaken by niche participants may be harbingers of the future.

Our analysis shows that the aim of transition management is not so much the realization of a specific transition described in a strategic vision: it may turn out that sufficient results are obtained via the route of system improvement or that problems are not as serious as believed. The aim of transition management is to provide an active contribution to the exploration of a transition that produces collective benefits in an open, explorative manner. Transition management is not geared towards gaining control of complexity, but much more towards dealing with complexity in a productive way.

Fortunately, elements of transition management are becoming part of modern governance of socio-technical change, at least in The Netherlands. Examples are the DTO programme on innovation by means of backcasting, the COOL dialogue on climate change, which organizes and facilitates discourse among stakeholders of change, NIDO, the national initiative for sustainable development, which organizes and facilitates new initiatives (so-called jump projects) for sustainable development, and the SER proposal for a Delta programme for sustainable energy. Transition management could be a next step in this kind of learning-oriented policy approach. Recently, the fourth Dutch national environmental programme also recognized the contribution of transition management to longer-term oriented change and development for climate protection and sustainable development.

Conclusion

Drawing on notions from technology dynamics and evolutionary theory, this chapter showed that incremental (step-by-step) socio-technical change dominates patterns of socio-technical development. The climate gains of this kind of incremental socio-technical change are far from enough for the emergence of the climate-neutral economy in 2050 as sketched in both the A and B visions in Chapter 6. The climate-neutral society requires the innovation of entire systems rather than the optimization of current systems. Where system optimization dominates technological-change processes, the climate-neutral society requires system innovation. The challenge for technology-oriented climate policies, therefore, is how to contribute to system innovation in a world changing and developing along patterns of system optimization. Drawing on evolutionary socio-technical change, the chapter suggested three changes.

Our first suggestion was that short-term technology policies should be based on a clear analysis of technological-change processes. Technology dynamics can help here, because a clear analysis of the dynamics of socio-technical change processes helps to develop policies that can influence (not forcefully steer) these dynamics in directions leading to system innovation. We suggested promising points of intervention for this kind of policy, being those points in technological-change processes offering potential for radical technological change. Our conclusion was that policies should focus on this potential for radical change (system innovation).

Secondly, we suggested an evolutionary-based set of policy instruments which is able to exploit the potential for system innovation. These kinds of instruments are able to initiate and facilitate the emergence of technological variation necessary for developing promising future technologies. These kinds of instruments should be combined with instruments coupling variation and selection and instruments safeguarding the market diffusion of technology by means of market selection. They can be either generic or specific in focus and content. Detailed analysis of the contingencies of technological change in specific settings should be decisive for the choice of types of instrument needed to exploit the potential for radical change in the specific setting: generic, specific or combinations of both.

Finally, we suggested transition management as a device for integrating short-term policies of technological change and innovation in the context of longer-term governance of system innovation.

Transition management is a societal 'tool', offering society as a whole a methodology for organizing and facilitating transitional change towards longer-term system innovation. Transition management suggests ways of organizing and structuring the process in such a way as to ensure optimal participation by all actors involved, and it offers suggestions for how to explore and develop future images of the system innovations that society fosters. Most importantly, transition management offers a device for short-term adaptive societal learning from the perspective of longer-term system innovation. Transition management, therefore, is highly valuable for modulating dynamics of socio-technical change. By means of adaptive learning, transition management contributes to the acceleration of innovation and socio-technical change because it organizes and mobilizes the actors and capacity society needs to achieve the climate-neutral economy in 2050. Transition management is, therefore, a societal (governance) challenge rather than a governmental challenge. Government has a special role and responsibility: acting in socio-technical change and directing the process from the perspective of longer-term transitional change.

These suggestions aim to contribute to the further internalization of climate protection needs in processes of socio-technical change. Working on the internalization of these needs is in fact the important thing that can and should be done, because we don't know what the future will bring in terms of innovation and change. We only know that we want this future world much more climate-friendly than today's world. Despite our ability to describe how we like the future world to be, either A or B or perhaps C, it is impossible to order its effectuation in 2050. We can only influence and redirect the processes that will lead for certain to *a* future state of the world, instead of the future state of the world. The only thing we know now is that we want this future world as climate-friendly as possible and, for that reason, we can and should more actively anticipate climate protection in innovation and technological change. This chapter suggested several general ways to achieve this, knowing that it is impossible to say whether they will lead to a future world as described by the A vision or a future world as described by the B vision.

For the short term, our evolutionary-based general suggestions translate in several more specified recommendations:

- Policies can influence technological change simply because technology results from human effort, but policy intervention contains no implicit guarantee of success (emission reduction) due to the contingency of actors and factors influencing technological change. Government has little or no control over these contingencies, so centralized general regulatory approaches make no sense.
- Policies always have to intervene in existing regimes and should build on promising ongoing developments in those regimes and move from there to the further consolidation of these developments.
- Since the economic logic of the market forces actors to focus on short-term technology improvements, economic incentives for CO_2 reduction (emissions trading, CO_2 taxes) make sense because they change the economic frame conditions. However, these shorter-term policies should be supplemented with an explicit effort by public and private actors to develop an agenda for system innovation. This calls for innovative policies fostering 'learning by doing' and 'doing by learning'.
- Transitional agendas, discourse (development rounds) for evaluation and adjusting goals and means can contribute to the design of transitional processes.
- Transitions take time. Rapid and forced developments may result in societal resistance and weaken the societal legitimacy of transitional targets and means.
- Supportive policies should never back just one horse (one technological alternative). Supportive policies should build on extensive knowledge of the technological and social ins and outs of systems to learn about the potential for a climate-neutral economy and to develop policies which can support the further rise and development of this potential.
- The climate change challenge demands action now. Policies should initiate change of technology in a climate-friendly

direction. Just waiting for change is no option in the timeframe of technological development.

- Technology policy should comprise both generic and specific instruments. Specific instruments, such as R&D programmes and technological experimentation, are needed for system innovation.
- Policies should not pursue one-dimensional goals, such as 80 per cent CO_2 reduction, because these kinds of goals do not reflect the wide range of changes needed for system innovation. Instead of one-dimensional goals, change processes should be guided by quality images of future developments. These images should be translated into differentiated intermediate and end goals, structuring the process towards system innovation.

❑❑❑

9

Opportunities for Change

Introduction

EFFORTS TO ACHIEVE A CLIMATE-NEUTRAL society will have important implications for society as a whole. The visions A and B, discussed in Chapter 6, give impressions of the drastic changes required to attain a reduction of greenhouse gas emissions by 80 per cent. The household sector constitutes an important part of society and with its consumption contributes substantially to the use of energy and the emissions of greenhouse gases. Looking at the implications of visions A and B for households, we discern several technical measures: high level of insulation of houses, new technologies for heating, cooling and other household activities and the use of renewable energy sources. The household consumption patterns – and more generally household behaviour—are not specifically analysed and described, although visions A and B do include some general sociocultural notions about consumption. Several projects have, however, addressed in detail the issue of reduction of energy use and of emission of greenhouse gases through changes in household consumption patterns.

Therefore, the main question we ask in this chapter is: is it possible to change household consumption patterns in a way that contributes to the attainment of a climate-neutral society? We assume that significant trend breaks are required with regard to household consumption patterns to reach the objective of a climate-neutral society in the year 2050. To achieve effective trend breaks some policy dilemmas need to be considered. The most relevant dilemmas are the multitude and diversity of actors involved (over 6 million households), the absence of the knowledge required to understand the 'global change' problem and to adopt effective measures to reduce emissions and the need for collaboration among industrial, administrative and institutional actors in developing strategies for the adoption of consumption patterns involving low greenhouse gas emissions.

To answer our question, we first discuss the present environmental performance of households in order to understand the main determinants of the greenhouse gas (GHG) emissions related to household consumption. Second, we explore the trend breaks in household consumption in the past 50 years in order to identify the main conditions determining the occurrence of trend breaks in the household sector. We then identify the options and constraints facing households in realizing radical reductions in emissions of greenhouse gases. From the results we can derive indications of the potential for achieving relevant trend breaks in the next 50 years. We will discuss four studies that have focused on possibilities for energy-saving consumption and, lastly, we will delineate opportunities for climate-relevant trend breaks.

However, the very first question is: will households survive? Von Schweitzer (1975) asked this question 25 years ago, although she posed the question for the year 2000. She assumed they would survive, and the essence of her reasoning was that households and families are more or less timeless. For centuries the basic functions of households have been to create a familiar space for everyday life and to fulfil people's daily needs. Children will always need a safe place to develop their character and skills and to socialize. Grown-ups need a place to recover, physically and mentally. There is no reason to assume that these personal and societal functions will become irrelevant in the next 50 years.

Because households are units of interacting and interdependent personalities, who have a common theme and goals, who have a commitment overtime and who share resources and living space (Hook and Paolucci, 1970; Aoberg, 2000), they form a specific context for decisions and activities in everyday life and consumption. Household consumption is therefore not simply a sum of individual consumption patterns. With their common resources like money and time, skills and knowledge, their appliances and other goods, as well as facilities and services from outside organizations, they produce an output: the level of living. Households try to bring their level of living in line with the standard of living. This is a set of norms and values about daily life, partly formed by values and norms in the outside world and partly by those of the household group, by which members formulate their goals, views and ideas about the way they live. The discrepancy between the standard of living and the level of living works as a controlling mechanism. When the level of living is not up to 'standard', household production will increase, another mix of resources will be used. If that is not possible, if the resources are insufficient to maintain the 'standard', due to a decrease in salary or capacity to work, for instance, households will adapt their standards. Money plays an important role, but is only one of the resources. Human resources are equally important in reaching the household goals.

Household consumption takes place in this system of everyday life. It is directed to the well-being of the household group but also of the individual members. Individuals are only part of the system. As Deacon and Firebaugh (1988) point out, there are individual sub-systems within the household system. These systems are flexible. As children grow older they acquire more power and contribute more.

In the two visions of society in 2050 presented in Chapter 6, assumptions were made about future human behaviour. It is interesting to relate these assumptions to the household setting. In Vision A people work hard, have a lot of freedom, seek self-fulfilment, are individualistic, have a progressive outlook and look for satisfaction in consumption. The market assumes parts of the household production, the activities for daily care. More meals are eaten in restaurants, there are more hairdressers and private childcare centres.

In Vision B the working week is less than 32 hours. People assign high values to nature, to social aspects and to justice in society; they do not seek individualistic self-fulfilment but strive for the development of a sustainable and social society, including prudent consumption, extensive public transport and collective arrangements for care of the elderly, the disabled and children.

To quote Von Schweitzer (1975) again, there are two dimensions to the way in which different households perform their functions. First, the economic-technical dimension of self-support ranges from a highly automated household at one extreme to a household that outsources as much as possible at the other. Second, the sociocultural dimension varies from individually determined needs to socially integrated mass consumption. We can position the two visions along these two perpendicular dimensions.

Vision A seems to represent the case of fulfilment of individually determined needs combined with an emphasis on outsourcing. Vision A creates ideal conditions for young, healthy, skilled people who are able to look after themselves, have a high income and can buy all they need. They have opportunities to spend time and money on holidays arid see the world. It is clear that this situation is less adapted to people with low incomes, people with physical or mental disabilities and to bringing up children.

Vision B represents the opposite end of the spectrum: self-support combined with socially integrated consumption. This vision envisages people adapted to social standards, but active in the private sphere. The young, healthy, skilled people will exercise a lot in gyms and do volunteer work in the community. Everyone who wants to work can find a place for their children in the crèche or kindergarten at a reasonable price. When children, elderly parents or neighbours are ill, it is possible to adapt work in a flexible way.

In both visions, household work will be easier than in the past. Appliances are user-friendly, materials in the house and clothing are of a high quality and easy to clean. With the market and privatization predominant in A, less time will be needed for household activities. Outsourcing becomes important but implies spending more money. Under B more time will be dedicated to household affairs and less money will be spent than under A. There is time for social activities and some household functions are performed collectively, but there is less emphasis on consumption of products that are now regarded as basic needs, such as daily meat consumption.

In a society according to Vision A, life within the household seems to play a limited role in society. If the general role of households is severely reduced, it becomes questionable whether society will survive without the production of some essential functions (*e.g.*, safety and security, education and socialization) presently delivered by the households creating at home a safe and stimulating place for the family members. Vision B is more idealistic than A, because it assumes the existence of values such as social responsibility and places less emphasis on individual material consumption, so the adherence to these values may offer better opportunities to attain a climate-neutral society. To arrive at the conditions of Vision B starting from the situation as it exists in 2000, norms and values in society will have to change substantially. Norms emphasizing a care and responsibility oriented society rather than a market-oriented society need to be developed.

Direct and indirect energy consumption by households

In the Dutch energy statistics the energy consumption by households constitutes only a minor part (20–25 per cent) of the total energy consumption in The Netherlands. These household energy consumption figures only reflect the use of electricity and natural gas by households. The figures do not include energy use for private and public transport and the energy use required for the production and distribution of goods and services acquired by households.

In our analysis we adopt the central assumption that all production takes place to serve consumption. So the energy consumption and greenhouse gas emissions of the various production sectors are attributed to the households in relation to the volume of goods and services consumed by Dutch households. With this method, we follow the energy use and emissions during the complete chain of Dutch household consumption: production, transport, trade and sale, use in households and disposal.

We distinguish two flows of energy and two corresponding sources of greenhouse gas emissions. The first is the direct energy requirement, which is the energy used by households in the form of energy carriers like natural gas, electricity and petrol. The second is the indirect energy requirement, which is the amount of energy used in the production and distribution of goods and services that can be attributed to households. The total energy requirement of households is the sum of the two.

Using this approach, the average household spending in The Netherlands was analysed and the average household energy requirement was calculated for the year 1990 (Biesiot and Moll, 1995).

Important spending categories in monetary terms are the house, food, education and recreation and medical care. The spending on direct energy requirements is very low.

In energy terms, the share of direct energy (petrol, electricity and heating) is substantial (46 per cent). The indirect energy requirements are found in the other spending categories (54 per cent). Food, education and recreation and household effects are the most important indirect energy categories.

The analysis is based on total household spending amounting to NLG 40,000 in 1990 (18,000 euro). The total energy requirement per household is 240 GJ/a. Around 60 per cent of The Netherlands' energy consumption is attributed in this way to households. Besides households, other relevant final energy consumers are the government and export sectors.

Additional statistical analyses demonstrated a strong correlation between calculated total energy requirement and the amount of household spending (elasticity coefficient of 0.80). They showed that there were large differences between households with an equal amount of household spending with respect to their total energy requirement (differences of up to 25 per cent from the average). Large differences also existed with regard to the energy requirements for consumption categories. For instance, the indirect energy requirement for food is substantial in relation to its share in total spending.

Trends in the total household energy requirement

The second half of the 20th century is very important for an analysis of present developments in household energy requirements. Economic growth and greater prosperity for households and families followed the period of post-war reconstruction in the 1950s. This resulted in a higher standard of living, better education and a desire for more possessions. The majority of households came to have more disposable income than they needed for the primary necessities of life. At the same time, family arrangements were more complex and life in the private sphere became more intensive. To ease the accompanying increase in household work, part of the money was spent on household appliances (Visser, 1969). The direct and the indirect energy requirements of households increased in line with these developments.

Van der Weal and Noorman (1998) have described the development of the household energy requirement for the period 1950–95. Before then, during the 1940s, Dutch households had to cope with a society where, due to the Second World War, technology developments for the domestic field were at a standstill and energy consumption was far from usual. For the whole period 1950–95 we observe that the share of direct energy in the total household energy requirement was below 50 per cent.

Substantial increases in demand for heat, electricity and motor fuel are observed in this period. After the 1979 direct energy consumption peak, we observe a remarkable decline of the direct energy requirement mainly caused by the substantial decline in demand for heat in the period 1980–95.

To take a closer look at these trends from a micro perspective, the use of natural gas and electricity by households is considered separately. After 1980, their paths diverge: after a period of near stability between 1980 and 1990, electricity consumption resumes its steady rise. Gas consumption fell substantially during the same period, before starting to rise again after 1990.

Electricity consumption in The Netherlands has been strongly influenced by the use of household appliances. The trend in the use of appliances was in turn constrained more by housing conditions than by availability of the appliances. Gas consumption increased tremendously after the discovery of a large natural gas field in the country during the early 1960s (see also Chapter 5 for a description of the introduction of natural gas in The Netherlands). The peak in consumption was followed by a period of energy saving, mainly due to improved insulation of houses. Both consumption patterns turned upwards again due to the economic boom of the 1990s. A careful analysis of the underlying determinants is required to recognize the most important trends and trend breaks.

Electricity

Electricity serves a wide range of household processes, mainly through the use of electrical appliances. Before 1960 Dutch households were equipped with at most one or two domestic appliances. Table 9.1 shows the development in the percentage of households with appliances — with a substantial annual electricity use — during the approximate period 1960–2000.

The washing machine was the most common appliance during the 19605, before being overtaken by the refrigerator around 1970. The popularity of the washing machine is not surprising since washing textiles by hand is a heavy job and such tasks are the first to be mechanized in households. Similar figures, only much earlier, are seen in the USA, where the washing machine is one of the first appliances bought by households. But there it was overtaken by the refrigerator around 1940. By 1950 four out of five American families possessed a refrigerator (Lomborg, 2001). The dish washer has slowly found its place in the home during the last 25 years, while the tumble drier became widely accepted in the 1990s. This lag in ownership of dish washers in Dutch households in relation to the tumble drier contrasts with developments in most other EU countries. Only the United Kingdom and Belgium show more or less comparable figures. In France, Germany and Austria, nearly 40 per cent of households owned dish washers by the middle of 1990 and the tumble drier lagged behind in those countries (Siderius *et al.*, 1995).

The penetration of these appliances in the households cannot solely explain the steep and steady rise in electricity use for so many years. To understand other possible factors behind it, one needs to look at how life was organized in the beginning of that period. Aspects of the daily life of housewives were investigated in 1964 in an extensive time budget study. Two facets are taken from that study: laundering and how facilities in the home limited the uptake of appliances.

As Table 9.1 shows, 75 per cent of households already had a washing machine in 1964. Of the remaining households which did not have a washing machine, most made use of laundering services or laundromats for some or all of their laundry. But doing the

Table 9.1 Electrical appliances in households from 1961 (%)

Appliance	1961[1]	1964[2]	1972[3]	1975[3]	1980[3]	1985[3]	1990[4]	1995[4]	1997[4]
Washing machine	67	75	86	86	89	88	88	92	93
Refrigerator	12	40	88	93	98	98	97	[5])	[5])
Freezer	–	–	17	30	46	48	53	61	65
Dish washer	–	–	6	8	11	8	10	20	26
Tumble drier	–	–	–	–	10	14	25	41	46

[1] Margriet (1962) Lezerskringonderzoek, Algemene analyse van de lezerskring. De Geillustreerde Pers N.V. Amsterdam.
[2] Philips (1966) De Nederlandse huisvrouw. Philips Netherlands N.V., Eindhoven.
[3] Statistics Netherlands (1992) Statistical Yearbook of The Netherlands 1992. CBS Publications, the Hague, Netherlands.
[4] Statistics Netherlands (2000) Statistical Yearbook of The Netherlands 2000. CBS Publications, the Hague, Netherlands.
[5] Nearly all households.

laundry was far different from the washing we know today. In 1964, nearly half of the washing machines were not connected to a tap and/or water outlet.

At the same time, synthetic fibres and the many types of non-crease textiles were appearing on the clothing and household textile market. They did not need intensive washing, but textiles from those materials had to be washed more frequently. Twenty years later, in 1984, almost every household had an (automatic) washing machine and only 4 per cent of them had to be moved from a room, corridor or garage to somewhere else to be used. By then, about 93 per cent of the machines were connected to a drain pipe and the other 7 per cent disposed of their water via a bath or shower room. Washing clothes was much more similar to how it is done in 2000.

Clean textiles had ceased to be a major headache (Philips, 1985). This affected quantities. The amounts washed increased from just over one kilogram of laundry *per capita* per week in the 1960s to over five kilograms of laundry per week at the beginning of the 1980s. And it did not stop there. The volume of washing *per capita* per week is presently well over 6 kilograms.

So we can identify the following factors determining the huge increase of energy use for washing: less washing by hand and a growing volume of laundry per person. Explanatory determinants for these factors are behavioural change related to changes in the types of textile in use and washing frequency, technological change in the washing process (automation and electrification of heating) and increased availability of these washing technologies (affordable prices and domestic infrastructure to connect and easily use these appliances).

After the Second World War, people generally developed a demand for mechanization. The electricity system was also ready to run the appliances. By 1956, the electricity companies had connected 90 per cent of the Dutch dwelling units to the grid. In the USA, domestic appliances were already being produced in large quantities. But appliances produced in the USA were very expensive by European standards and were difficult to acquire for Dutch households. The mechanization of Dutch households only became feasible when European manufacturers started making relatively cheap domestic appliances in larger volumes.

In The Netherlands, the possession of household appliances was also influenced by another important phenomenon. Albert Heijn, one of the major grocery chains, wanted to expand its market by selling cooled food products, like the grocers in the USA. However, with such a low penetration of refrigerators in The Netherlands, people could not properly store the cooled products. Therefore in 1962 Albert Heijn started a very successful trading-stamp scheme which allowed households to save stamps to buy refrigerators and other domestic appliances at a discount. The scheme lasted for more than 10 years, and when demand for domestic appliances was saturated, the campaign shifted to appliances for do-it-yourself activities and personal care.

However, as several time budget studies of households have shown, introduction of technology may lead to time saving in individual processes, but it does not do so for whole household tasks. Although mechanization did not help to save time, it provided ease of work, comfort and opportunities to answer the greater demands of a more complex living standard. Households were better able to enjoy a higher standard of living (Groot-Marcus, 1984).

Technology improves the quality of life of the person who performs household tasks and of the other members of the household at the same time.

Because of constraints in the domestic infrastructure and in the supply of appliances, household consumption of electricity did not explode but increased gradually in the second half of the 20th century. The facilities for washing became more and more sophisticated, as did the domestic infrastructure for showering and bathing.

What is important in the context of washing clothes—and also with regard to showering and bathing—are the values and norms for hygiene and cleanliness. Quality standards in households are not fixed. When a certain level of quality is reached and only a moderate effort has been needed to get there because of technological developments, the goals are set anew. There seems to be no limit to growth. This is apparent in the cases of bathing and washing. It is not the necessity of hygiene that dictates the frequency of bathing, showering and washing, but the change in norms and the ease with which they can be met.

Housing

Because of the housing shortage after the Second World War, which lasted well into the 1960s, houses were too small and inflexible to accommodate an expanding number of appliances per family (Visser, 1969). Around 1965, the minimum requirements for new houses were raised, and towards the end of the 1980s the minimum floor space was increased again. Broadly speaking, the houses of the 1990s are about 40 per cent larger than 40 years earlier, while the number of persons per household has declined at the same pace (from nearly 4 persons per dwelling to 2.45 in 1994). Today, nearly all houses have a shower or bathroom, and sometimes more than one, and they are better equipped. For one room the government did not raise the minimum dimensions: the kitchen. In 1965 kitchens could be built with just 5 m^2 of floor space, and that was still the case in 1990. Kitchens with the minimum dimensions are designed for two appliances, an oven and a refrigerator (van der Meer, 1989). Scarcity of space may explain the slow acceptance of dish washers in Dutch homes and the preference for buying a tumble drier, which can be placed elsewhere.

Heating

Heating has a rather different history than electricity use. Dutch houses switched almost completely from heating with coal stoves to central heating with natural gas in less than 20 years. This changed life quite dramatically in the 1960s. The availability of cheap natural gas led to government campaigns in the 1960 to persuade households to use gas for cooking and heating. Central heating was also promoted, resulting in 30 per cent of houses being centrally heated by 1971. By 1984, 75 per cent of houses were centrally heated by natural gas. Most of the 13 per cent of houses that were not centrally heated in 1995 used gas stoves in more than one room.

After the massive switch to central heating, the consumption of gas became exorbitantly high. At the peak of the domestic gas consumption, around 1980, the required energy per household went up to 70 GJ/a. Stimulated by an extensive publicity campaign and government subsidies, in the 1980s people started to insulate their houses. In the 1990s, annual energy use declined to about 40 GJ per 2–5-person household. This amounts to 16–18 GJ per person, which, given that the total energy requirement of a four-person household for heating, cooking and lighting was about 25 GJ a year in the middle of the 1960s, was three times more than 35 years earlier, a substantial rise in energy consumption.

We can identify the following factors that determine the increase in gas consumption by households. The easy availability of natural gas triggered a lot of behavioural changes including a longer heating season, an increase in the average indoor temperature, a substantial rise in the number of rooms heated in the house and an increase in the use of heated water for bathing and showering. The expansion of thermal insulation in houses reversed—especially in the period of high energy prices—the rising trend of natural gas consumption.

We can also see that central heating created opportunities to exploit much more living space. This development cannot be reversed. Dutch households cannot live their lives in one room any more. Current social norms dictate that every member of a household has the right to personal development and privacy.

Trends and trend breaks with regard to direct energy consumption in the 1960–95 period

Consumer behaviour is determined by three main factors: individual and household needs, opportunities (external facilitating conditions) and abilities (internal capacities) (see Gatersleben and Vlek, 1998 for a description of this explanatory model).

The trends in gas and electricity consumption are related to changes in these factors. First, trends relating to 'opportunities' (*i.e.*, the connection of households to the natural-gas pipeline system and to the electricity grid and the availability of household appliances and heating systems) partly explain the substantial increase in the direct household energy use. But the increase in the 'abilities' of households (related to the rise of personal income and the increase of interior space per household member) is also important in explaining the rise of direct household energy use. Another interesting factor is the observed feedback from the 'opportunities' and 'abilities' factors to the 'needs' factor. The needs also changed during this period: a need for privacy related to the increased availability of (heated) space, a need for higher standards of hygiene and cleanliness because of the opportunities to achieve such higher standards. So with regard to the direct energy use in households we observe a co-evolutionary process: economic changes, changes in infrastructure and sociocultural changes (affected by the changes in individual and household needs) influence each other in an intricate way, resulting in the steeply rising trend of direct energy consumption in the 1960–80 period. Some (temporary) trend breaks after 1980 are also explained by these factors. The economic recession during the 1980s temporarily limited the growth of electricity use (through decreased financial abilities). The availability of newly developed energy-saving technologies (insulation, highly efficient heating appliances) curbed the use of natural gas.

These and other social trends determine the elements of the direct and indirect household energy requirement not discussed so far. The greater use of motor fuels can be analysed in terms of a similar co-evolutionary perspective. A closer inspection of the determinants of the changes with regard to the indirect energy requirements yields interesting insights: relative increases in the categories 'leisure and education', 'personal care' and 'household furniture and decoration'. These findings are related to social trends like individualization, emancipation, emphasis on personal development and increase of leisure time.

Notwithstanding a general increase in environmental awareness in society, its influence on direct and indirect household energy consumption appears to be marginal. The energy use for heating may have fundamentally declined in part for environmental reasons, but although the use of electricity, motor fuel and the indirect energy use declined for some years because of an economic recession, they resumed a steady growth path in the succeeding period of economic recovery.

Feasibility of substantial greenhouse gas emission reductions by households?

First, this chapter presented an accounting for the total household energy requirement. These accounts demonstrated that a substantial portion of Dutch energy consumption should be allocated to households. The previous section described and discussed some trends that led to a substantial increase in household energy requirements in the decades after the Second World War. Substantial reductions are also required within households to meet the long-term emission targets for greenhouse gases. Based on this notion, several research programmes have been carried out to assess the potential for reduction of energy use and/or greenhouse gas emissions through changes in households. We will discuss four of these studies: the Perspective project, the GreenHouse programme, the HOMES programme and the SusHouse programme. These studies adopted an integrated approach, addressing the (macro-level) relationships between sectors and the relationship between production and consumption.

Several investigations identify options to reduce the household energy requirements and the related greenhouse gas emissions using the distinction between direct and indirect energy requirement.

The *direct* energy requirements of households can be reduced along two lines. The first line, a technological line, is directed at technological improvement of household appliances (low-energy refrigerators, high-efficiency heaters and boilers), improvement of the physical infrastructure (increasing efficiency in electricity production, reducing the need for physical transport), and investments to save on fossil energy (insulation of dwellings, use of waste-heat sources, use of renewable energy). The second line, a behavioural line, is directed at bringing about changes in household behaviour leading to reduced use of appliances and lighting, lower temperatures in space heating and water heating (for clothes washing), and in possession of fewer appliances.

The *indirect* energy requirements of households can be reduced along the same lines. The technological line is targeted at improving energy efficiency and saving fossil energy in the production and trade sectors; the behavioural line is aimed at changing the consumption pattern into a low energy-intensity pattern, such as buying green products, buying products with a low energy intensity and substitution between spending categories.

Perspective project

The 'Perspective' project involved a practical study of the possibility of reducing energy consumption through information-induced behavioural change of consumption patterns (see Schmidt and Postma, 1999). Twelve households participated in a two-year field experiment. Their target was to reduce direct and indirect energy consumption by 40 per cent, notwithstanding the 20 per cent rise in income they were awarded for meeting the reduction target. Households had to fill in weekly lists concerning their energy use and shopping and received direct feedback through informative software on the energy consequences of their behaviour. In addition, tutors gave intensive counselling to the households during the experiment.

The households almost succeeded in meeting the 40 per cent reduction target. Most of the reduction related to the indirect energy requirement. The direct energy requirement was reduced by energy-saving behaviour and by replacing old appliances with more energy-efficient ones. Important changes observed in consumption were:

- in creased purchase of expensive labour-intensive goods (designer goods, handmade goods and works of art);

- increased purchase of high-quality and durable goods (clothes, shoes and furniture) and extension of the lifetime by repair;
- alteration of diets (less meat, biologically produced food, and seasonal vegetables instead of vegetables produced in greenhouses);
- changes in the mobility pattern, including mobility reduction (short-distance holiday locations and cycling for commuting);
- increased purchase of services (domestic help, visiting restaurants and education, beauty services).

Lifestyle changes arising from this experiment were generally regarded as positive. The purchase of high-quality goods and personal services formed a very attractive ingredient of the new lifestyle. The participants initially found the constant attention to energy while shopping to be a burden, but became accustomed to this aspect of the experiment later. The attitude to the restriction of mobility was negative throughout the experimental period.

The lifestyle of the participants was also analysed one and half years after the period of the experiment in a follow-up survey. Changes in diet had been maintained, partly to almost entirely. Changes with respect to purchases of high-quality goods and personal services were partially maintained. Most of the changes regarding transportation and holidays had been abandoned. In addition, the direct energy consumption had increased moderately. Two setbacks can explain these results. The discontinuation of the financial contribution limited the possibility of spending money on quality goods and services. The discontinuation of control and feedback eased the pressure to maintain the discipline required for adherence to an energy-saving lifestyle. The participants gladly switched back to an intensive mobility pattern after the experimental period.

GreenHouse programme

The GreenHouse programme (see Moll, 2000; Uitdenbogerd *et al.*, 1998; Nonhebel and Moll, 2001) was designed to reach an integral assessment of the medium-term potential for reducing the energy requirement and greenhouse gas emissions of households.

This programme defined a comprehensive package of reduction options that are potentially feasible in the period 2000–20. With a constant average household budget, the potential reduction is 54 per cent in energy terms and about 50 per cent in terms of greenhouse gas emissions. Given the assumed increase in incomes (2.7 per cent annual growth rate of GDP) and energy efficiency (1.7 per cent annual decrease in energy intensity) according to the European co-ordination scenario of the CPB, the absolute reductions achieved with the 'GreenHouse package' compared to the 1995 level are about 10 per cent for energy and greenhouse gas emissions in 2020 (see Wilting *et al.*, 1999).

The package consists of technical improvements aimed at reduction of direct energy use (20 per cent), behavioural measures aimed at reduction of direct energy use (3 per cent), technical improvements in the production and distribution of goods and services aimed at reduction of indirect energy use (26 per cent) and changes in the consumption pattern effecting changes in the indirect energy requirement (5 per cent). About half of the reduction could be achieved by measures taken in the production sectors (supply-side options); the other half should be achieved by changes within the households (demand-side options).

This package is largely based on technical measures, and the role of behavioural change seems negligible. Of course, some technical measures also have a behavioural component: the acquisition of an energy-efficient refrigerator is also an aspect of behaviour, depending on the price, the availability, information and so on.

Complementary to the package design, a survey was conducted among households to investigate their knowledge about and the feasibility and acceptability of the various reduction options. According to the survey, the most well-known options were energy-saving light bulbs, driving less in a car, using a bicycle, lowering average room temperature, water-saving shower heads and washing at lower temperatures. The set of less-well-known options included replacing frozen or fresh imported vegetables with preserved vegetables.

The acceptability of the reduction options in the survey was measured in two ways. For each option, the respondents were first asked to give an acceptance percentage ranging from 0 to 100 per cent. At the end of the survey the respondents were then asked to choose four reduction options out of 32 that they would like to try and four options they were most reluctant about. Because the survey provided many details regarding the consequences of the options, respondents could estimate the implications for their own situation quite well.

The most popular reduction options were energy-saving light bulbs, energy-efficient household appliances, replacement of cut flowers with plants or other presents, water-saving shower heads, replacement of greenhouse/imported vegetables with preserved vegetables. The less popular options were no car, no freezer, no drier and greater use of public transport.

Although options were designed in such a way that implementation would have a limited effect on household practices, the willingness of households to implement the suggested options was very limited. Only options with barely any effect on household practices (such as the use of energy-saving light bulbs, purchase of an energy-efficient washing machine, etc.) could be interpreted as feasible under present conditions. There was one exception, involving the composition of the menu. Changing to less greenhouse gas-intensive vegetables (seasonal products) seemed a realistic option. Also striking was the fact that none of the options that affected mobility proved feasible.

The survey results demonstrate that, without any additional measures in infrastructure etc., on average only a part of the demand-side potential for reduction of greenhouse gas emissions is currently feasible: an estimated 5–10 per cent of total household energy requirements. This estimate is much lower than the indication (around 25–30 per cent) based on technical analysis of the envisaged reduction package (Uitdenbogerd *et al.*, 1998).

HOMES programme

The HOMES programme (Noorman and Schoot Uiterkamp, 1998) addressed the issue of sustainable household metabolism. This issue was elaborated in some aspects of the household consumption pattern: space and water heating, the acquisition and use of (electrical) appliances and transport. A multidisciplinary perspective was followed, including socio-psychological, spatial planning, policy and energy/environmental analysis. From each of the disciplinary perspectives, the programme presented a vision of the possibilities and constraints with regard to realizing a more sustainable household consumption pattern. Some important findings are presented below.

The energy analysis focused mainly on direct household energy use. A scenario approach demonstrated a potential 40 per cent reduction of direct household energy requirements in the Towards Sustainability scenario, and a 10 per cent increase of direct household energy requirement in the Business-as-Usual scenario (Van der Weal and Moll, 2001).

On the basis of a survey, the socio-psychological analysis demonstrated that individual respondents may accept moderate reductions (10–15 per cent) of their total energy requirements in order to achieve a more sustainable society. Respondents also believed that reductions of up to approximately 25 per cent will not have a major effect on their quality of life (Gatersleben, 2001).

The spatial planning analysis addressed the relationship between forms of urbanization and household energy requirements. It emerged that the position of the urban centre in its surroundings is an important determinant of a household's mobility in practice. For a regional centre within a largely rural area, compact urban development demonstrates the emergence of favourable mobility patterns. In other areas that already have high-density urban centres (the so-called urban fields), new compact urban areas will stimulate mobility because of the large diversity of spatial relations in that case (Van Diepen, 2000).

The policy analysis addressed the issue of steering households toward a more sustainable consumption pattern. Here, the production-consumption chain is also very important. All actors (producers, retailers, consumers) in that chain should be targeted. Steering producers and other economic actors is easier than steering the consumer because of the wide diversity of the latter group and also because of the remoteness of the government from the consumers. Indirect mechanisms (through economic actors) and unilateral mechanisms (taxation and information) were recommended to steer consumers effectively (Ligteringen, 1999).

SusHouse programme

The SusHouse programme analysed the long-term feasibility of sustainable household practices (for food, see Quist, 2000; Dekker, 1999). Potentially sustainable scenarios were derived for some important household activities from a mixture of forecasting and backcasting techniques. The forecasting process identified major economic and sociocultural trends for some household activities. Based on these trends, some long-term scenarios were constructed. In the second stage, the consistency of these scenarios with sustainability goals was analysed, with the aim of assessing social and technical conditions crucial to meeting the sustainability goals within these scenarios. In the third stage, the backcasting, ways of reaching a future state corresponding with the sustainable scenarios were explored.

With regard to food, the SusHouse project produced three greatly differing scenarios: the intelligent kitchen, the 'superrant', and local food production.

1. Acceleration and intelligent high-technology cooking: Households are characterized by high-tech, convenience, do-it-yourself, moderate use of services and acceleration of the pace of life, especially during the working week. There are no fixed meal times, but meal periods during which one can have meals depending on working hours and leisure activities. Environmental gains arise from management optimization in such a system and the use of sustainable ingredients.
2. The neighbourhood food centre (also named the superrant): In a compact city people can go to centres for a meal, grocery shopping, to purchase take-away meals and eat together for different prices. This neighbourhood food centre is visited by a wide variety of households including single people, families and elderly people. Environmental benefits stem from local upscaling, sustainably-grown foods and increasing the service component.
3. Local green menu through green consumer demand: Households buy and eat seasonal food grown locally and

purchased at local shops, small supermarkets and from the grower as 'fresh' unprocessed ingredients. Imported products are expensive due to the incorporation of environmental costs. The coupling between food and origin is enhanced and all the cooking is done in the kitchen at home. Environmental benefits arise from sustainable, seasonal and locally grown foods, accompanied by the minimization of transportation and processing.

Assessing these scenarios against sustainability goals, for instance with regard to greenhouse gas emissions, with each scenario there are major difficulties in meeting these goals, and additional far-reaching assumptions are required, for instance, as regards efficiency of appliances and personal and goods transport in the first two scenarios and with regard to the personal diet in the third one.

Reviewing these four partly complementary studies, we can draw some conclusions about the reduction potential related to household consumption.

The Perspective project shows a substantial potential for behavioural change, effectively resulting in substantial reductions even in the case of growing personal income. Comparable results are found with Eco-teams, using information transfer, self-monitoring and social stimulation (see also Chapter 7). But we should be aware of the considerable initial effort put in by the households participating in the Perspective project and by Eco-team members. The induced behavioural change will also only be partially maintained without continuation of the special attention given to the people and constant flows of feedback and information.

The GreenHouse and HOMES programmes followed a mixed approach to reducing greenhouse gas emissions: technological development and implementation targeted at energy efficiency and behavioural change. In the event of growing personal income, the gains of this combination of approaches are severely reduced. Nor is there a high degree of acceptability for the assumed behavioural change for most of the reduction options. The HOMES programme demonstrates that additional approaches—beyond the technological and behavioural approaches—such as an institutional or a policy approach and a spatial planning approach may offer better prospects for far-reaching reductions. Changing the institutional and physical infrastructure around the households may accommodate more environmentally friendly and energy-saving behaviour than the present social and physical environment of households. An example is provided by the study about deprivation of the rural poor due to their inability to travel in the UK. To create sufficient possibilities for them for a decent way of everyday living, a policy package is needed consisting of an adequate range of services. First, facilities at local level need to be supported by the village community or local authorities (grocery shops, library, banks, medical services); second, better public transport, supported by national regulations as well as by local authorities; and third, compensation in traveling costs for individuals with a low income. Better public transport will also lure car owners out of their cars. The challenge here is to integrate the measures implemented by different actors (Boardman, 1999).

The scenarios of the SusHouse project assume new sociocultural standards as well as a new infrastructure. To achieve a SusHouse scenario, trend breaks are required because of the large differences between the scenarios and present-day standards and infrastructure. The results of an evaluation of the SusHouse scenarios again demonstrate the difficulties of realizing absolute reductions in a world of growing consumption.

The overview of these programmes may also provide insight into the characteristics of the options at the household level that would yield reductions of greenhouse gas emissions and are acceptable to a substantial share of households. Technical options based on increased efficiency of appliances and light bulbs are generally acceptable. Intensification of energy-saving behavioural patterns—already existing in the household—with regard to the use of appliances is accepted in many cases. With regard to indirect energy use, many substitution options (giving other types of presents, some changes in food consumption) are also easily accepted when enough information is available. The acceptance of some other options is very low, such as the abandonment of appliances, the reduction of car use and the restriction of choice with regard to holiday destination. Characteristics of such options are that they are perceived to affect the quality of life and to limit the households' ability to meet their standard of living.

Voluntary moderation of consumption, although in principle a very effective way of reducing emissions of greenhouse gases, is not adequately addressed in these programmes. The Perspective project was designed to shift energy and material consumption to high-quality and energy-extensive services, but did not emphasize moderation of consumption per se. Some options in the GreenHouse and HOMES programmes addressed reduced possession and use of appliances but not general moderation of consumption. Voluntary moderation of consumption seems to be contrary to the present-day economic growth trends and sociocultural norms. So, for the time being we cannot expect much success from the stimulation of voluntary moderation of consumption.

With respect to the dilemmas discussed in Chapter 7, some remarks can be made about the feasibility of substantial greenhouse gas reductions by households. Not all dilemmas are relevant at this level.

The multitude of actors in the household sector poses a severe steering problem. According to Ligteringen (1999), indirect (unilateral) strategies are required for steering in such a case. The 'centrally-directed technological' trajectory, with a small number of actors involved in decision-making and implementation, is potentially a satisfactory strategy for reaching reduction targets. However, this strategy also poses the risk of the rebound effect, spurring the growth of consumption. Additional, indirect strategies should be included, aimed at securing changes in the social and physical infrastructure that facilitate household behaviour which leads to reduction of greenhouse gas emissions.

It is also clear that by contrast with other sectors in society the household sector is characterized by a diversity of small actors with no expertise in the energy field. The diversity of ideas, goals and contexts that exist in everyday life poses severe problems for efforts to get people to commit to a coherent and broadly accepted reduction strategy. Moreover, besides the layman's lack of knowledge of ways to change behaviour, everyday life nowadays is quite complex and dependent on arrangements made in different fields of society. In order to reach the current level of living and consumption, in the course of time many supporting organizations and infrastructures have been established and institutionalized. It is very difficult for those underneath to escape from this entrenched way of living. De-institutionalization will be more feasible if the lead comes from other social levels, such as local government.

The dilemmas about creating awareness and building up support through integrated sustainable development strategies are also relevant because of the multitude of actors and because of the lack of common knowledge and understanding about the climate issue as well as the sustainable development implications. The Perspective project and the 'Eco-teams' demonstrate the possibility of resolving these dilemmas in specific settings.

The co-benefit approach mentioned in the sixth dilemma would be more successful in the case of households. The first concern of a household will always be to match its level of living to its chosen standard of living with regard to the satisfaction of everyday needs of the household members. Households have developed skills, structured their lives, formed ideas and habits and acquired resources to reach this goal. If this goal can be achieved better or more easily by measures with a co-benefit for the environment, they will happily accept them. All programmes discussed above demonstrated much higher acceptance of options with co-benefit potential.

We can conclude with the following findings. The contemporary trends in society and the economy will not generate autonomous reductions of the greenhouse gas emissions related to household consumption. The growth of personal income combined with a low acceptance of energy-saving behaviour will nullify the expected efficiency gains in production and consumption technology. On the other hand, it has been demonstrated that substantial reductions through changes in the household sector are feasible under favourable conditions.

Opportunities for climate-relevant trend breaks in household consumption

The research with regard to sustainable household consumption demonstrates that a contribution by households to a societal development in accordance with Vision A or Vision B is doubtful.

A development according to Vision A corresponds to present-day standards in households. These standards largely determine the current trends in society. Extrapolation of these trends does not produce climate-neutral household consumption scenarios. The additional technological trend breaks in Vision A—over and above common Business-as-Usual scenarios—will probably spur household consumption further because this vision will not uncouple technological development and consumption growth. In this vision the consumer is not held responsible for his/her consumption. How can we expect the market parties and the low-profile government to take full responsibility for the reductions? A more fundamental question is who will take responsibility for the reproduction of the human population if individual short-term motives are dominant and the values necessary to create and maintain family life are almost absent. And how will the social infrastructure be reproduced if the investments required are only profitable in the long term?

A development according to Vision B seems to be more effective than Vision A to achieve substantial long-term reductions and to facilitate climate-neutral household consumption. However, the social values, responsibility and moderation assumed in that vision are fairly inconsistent with present social values and aspirations. So a social development in accordance with this vision will lack general acceptance among the present population. Moreover, the contemporary physical, institutional and social infrastructure does not accommodate a responsible and moderate lifestyle. So radical social and institutional trend breaks are required because the changing of social values and the renovation of infrastructure both require a long time span and constant intensive efforts.

The achievement of climate-neutral household consumption patterns therefore requires trend breaks at the technological as well as the social and institutional levels. In other words, a co-evolution of parts of Visions A and B. A co-evolutionary development of technical and social systems has occurred before in The Netherlands. The description of trends in household consumption demonstrated the simultaneous occurrence of far-reaching trend breaks at the technical and the social level. Households did not intentionally increase their energy consumption, because energy had a price, and in part of that period even a very high price. The increase in household energy use was mainly due to the implementation and growth of new technical systems and radical social changes, such as the construction of the natural-gas pipeline system, new standards for the building of houses, new policies on spatial and road infrastructure planning, and the general democratization of society and the emancipation of women. A deliberate design of future technical systems and institutional and social measures with the indirect effect of reducing energy use may prepare the conditions for climate-neutral consumption patterns by households. Parts of Visions A and B can be used to develop this approach.

Present-day societies are pluriform, made up of groups of people with different cultures, needs, ideas and standards. Ideas and needs also change during a person's life. Nor are countries and regions uniform; there will always be regions where people live close together and more rural areas. Positive appreciation of this pluriformity should also function as an antidote to the mechanism of social comparison which has stimulated growth of consumption in the past.

Vision A focuses on an important segment in present-day society: young, active people with a reasonable income living in smaller or larger towns. Climate-neutral consumption patterns imply that these groups in particular should give great attention to energy efficiency in their consumption and to the use of renewable sources in production and consumption technology. Adaptation will not be difficult if it can be advertised as typical consumption for their peer group. The implementation of these technologies is strengthened and the risk of rebound behaviour is mitigated by the implementation of regulatory eco-taxes on producer and consumer energy consumption. Producers are allowed to include the taxes in the consumer price. In that way, the prices of direct and indirect energy consumption are equalized at the consumer level. Full energy pricing may steer consumers to more responsible behaviour regarding energy use.

Vision B mainly addresses groups of people who are more concerned for their surroundings and the networks around them. These households should be informed and stimulated to behave in a responsible way. Emphasis should be given to the dissemination of information about the need for and advantages of new technologies for energy and CO_2 reduction. The development of alternative energy sources should have broad support among these groups. Private individuals, companies and households should see the necessity of such developments and people who have money should act as innovators, buying and trying out newly developed photovoltaic systems, sun collectors or windmills. Information must be made available in a variety of ways (labels on products, individual information and calculation software, tutoring groups to change behaviour) and on multiple levels (education, community centres, work place). In addition, incentives should be created to buy energy-saving products, to experiment with energy-friendly behaviour and to maintain and improve environmental lifestyles.

To achieve climate-neutral consumption patterns it is very important to give sufficient attention to the structures in society that were formed at other times with different goals, norms and values as well as technology in the field of energy production and use. The physical, social and cultural infrastructure should be deliberately reconstructed in such a way that responsible behaviour is facilitated and undesirable behaviour is curtailed. In many cases, the co-production strategy should be worked out. Restructuring the parking and traffic infrastructure of neighbourhoods and a deliberate choice of location for social facility centres may induce changes in the modal split together with an agreeable local environment and an easing of the time-stress on households. The deliberate design of environmentally-friendly service facilities may enhance the quality of life as well as reducing energy use.

Many trend breaks are necessary to achieve climate-neutral consumption patterns. A co-evolutionary process is required to realize radical emission reductions. As we argued in this chapter, consumer behaviour is determined by needs, opportunities and abilities. Changing needs is vital for a change in behaviour, but it does not have to start there. It appears that in the last 50 years changes in needs have been triggered by developments in the opportunities and abilities of households. It must be remembered that the co-evolutionary process seen in the last century was aimed at growth rather than reduction of material consumption and energy use. But an emphasis on reduction is not appealing to people, so attention must be drawn to the benefits that will ensue. In the anterior growth path, technology took the lead and the political and social system followed. Nowadays there are signs that the technology-push is losing its influence and that technology is developed more and more in interaction with the societal and (maybe) environmental demand. In the future reduction path, government policy should take the lead because of the inability of technological and social systems to initiate processes that will achieve absolute reductions.

What strategies should be used to resolve the dilemmas and push households in the direction of a climate-neutral policy? Based on the insight of the research programmes on more sustainable consumption, we can present three policy approaches that should be adopted together.

First, make available general information about the climate system and the role of energy consumption as well as information about specific energy-saving options at technical and behavioural level. This information must reach every consumer who might be inclined to consume with climate responsibility. Various information and implementation strategies should be followed to reach interested households: by Internet and software-based tutoring directed at individuals, dedicated information for city and neighbourhood groups and the formation of local teams practicing climate-neutral consumption, the promotion of energy-efficient consumption in magazines of grocery retailers, and so on. In this it is important that elements of the energy-saving consumption patterns should be associated as far as possible with other values, such as health, quality and innovativeness.

Second, national and local government should stimulate and implement indirect strategies to change household behaviour. Direct strategies are difficult and ineffective because of the enormous number of households and the pluriformity of the household sector. Some indirect strategies should address other actors, such as producers, and retailers as well as institutional organizations representing segments of the household sector. Other indirect strategies could influence market conditions with a view to achieving reductions in direct and indirect energy requirements. Taxation of energy and of consumer goods and setting standards with regard to the direct and indirect energy requirements of products are potentially effective ways of steering consumption.

Third, local and national government should adapt the general conditions for household consumption in a climate-neutral direction. The physical infrastructure (houses, roads and transport systems, energy systems) should be renovated and restructured to facilitate an agreeable lifestyle and climate-neutral behaviour. The co-benefit approach is very important. Solutions for local problems with regard to quality of life should be linked to the global climate issue. The production and trade sectors should be encouraged to shift the present emphasis on possession of goods by consumers to an emphasis on the delivery of services and to take full responsibility for the product life cycle. The development of climate-neutral product-service systems may also generate co-benefits.

There are opportunities for the emergence of a climate-neutral consumption pattern by households, but in most cases more is needed than goodwill from households alone. The achievement of climate-neutral consumption presents a major challenge, for society as a whole to strive for this goal, and for national and local government to develop strategies and secure support for organizing society in an environmentally sound way.

❑❑❑

10

The Role of Local Authorities in a Transition towards a Climate-Neutral World

Introduction

ALTHOUGH CLIMATE CHANGE is first and foremost a global environmental problem, it requires action at all levels of government. Primary responsibility for the implementation of international commitments, like the Kyoto agreements, lies with the national governments that signed them. The European Commission is a separate signatory. In Europe, these two levels of government are crucial for climate change policy because they set the appropriate climate change strategies.

However, it is clear that climate change has a local dimension as well. This chapter discusses the role of local authorities in a transition towards a climate-neutral society. Literature and policy practice throw up several general reasons why local authorities are important to climate policy. First of all, as the layer of government that is closest to the citizen, local authorities can play a major part in creating a degree of support for climate policy. Secondly, local authorities are responsible for local activities that result in greenhouse gas emissions, ranging from public lighting to the energy consumed in public buildings. And thirdly, they have responsibility in many areas of policy that are significant for climate policy, such as town and country planning, construction and transport. These responsibilities are not just very important for their own, local climate policy; they are also significant for the role of the local council as co-executive of national and international climate policy.

Given their position as the level of government closest to the citizen and their limited, but crucial, competence in important areas of climate policy, in this chapter we pose the question: *what can and should local authorities do to contribute to an 80 per cent greenhouse gas emission reduction?*

The answer to this question depends on the particular national, political and institutional context in which these local authorities have to operate. The structural features of this context play a significant role in how (local) climate policy is adopted and then put into effect. The substance of local climate policy depends on the specific nature of the community in question, such as its geography, demography, economy, society and culture (compare Lafferty and Eckerberg, 1998). We call the framework in which a local authority has to operate its *playing field.* For the answer to the question of what the role of local authorities can be in a transition towards a climate-neutral society, it is crucial to realize that this playing field differs between countries and can change overtime.

In the first place, municipalities in different countries around the world have different playing fields as regards their powers to steer and their economic possibilities to implement measures. Given these differences, one cannot simply conclude what municipalities should do from international documents. Supranational initiatives leave considerable room for cross-national variation to reflect the diversity in institutional and political local context.

Second, the playing field for local authorities is not static. For instance, the two visions of how emissions could be reduced by 80 per cent/described in Chapter 6 of this book, contain different assumptions about the role of local government in the future (see below).

Because of the specific characteristics of countries, it is difficult to define a general playing field for local authorities and, perhaps even more importantly, to define what needs to be changed in this playing field to optimize the contribution of local authorities to a trend break towards a climate-neutral society. There are two options that could potentially enlarge the municipal contribution. The first option is to increase the role of local authorities by expanding their competence and playing field. The second is to use the municipal playing field as advantageously as possible.

Option one, an enlarged role for local authorities, assumes a change in the role of local authorities in climate policy. In the next section we will discuss the role of local authorities in climate policy, the factors that determine this role and the potential changes in this role. Option two, optimizing the use of the existing playing-field, is specific to each country and local authority. We will discuss this trend break using The Netherlands as an example. Then we will discuss the role of local climate policy in The Netherlands, the local climate options and their total potential for greenhouse gas emission reduction and the use of their playing field by the Dutch municipalities.

Finally, in the last section of this chapter we will draw some conclusions about the potential role of local authorities in a transition towards a climate-neutral society.

THE ROLE OF LOCAL AUTHORITIES

Factors determining the role of local authorities in climate policy

We can address the role of local authorities in a change towards a climate-neutral society from both a normative and an empirical perspective. Empirically, we can ask what is the role of local authorities in climate policy and what factors determine this role. Normatively, we can ask what this role should be and according to what criteria.

Climate change is a classic example of multi-level decision-making (Collier, 1997). At the highest level of decision-making we find the United Nations, with the Framework Convention on Climate Change (UN FCCC) and the Kyoto Protocol. At the supranational level, in Europe the European Union influences climate change through its energy-efficiency, renewable energy and energy R&D activities, the EU structural funds, the common agricultural policy and the internal market. At the national level, national governments bear responsibility for economic, energy and transport policies.

Although there are many differences between countries, local authorities are generally responsible for local transport planning, spatial and development planning and energy management. Last but not least, there is the level of the citizens and communities where decisions are made about the ultimate use of energy and choices are made about transport and the use of renewable energy.

In general, the powers of local authorities with respect to energy and climate matters vary greatly from one country to another. By analogy with Local Agenda 21 implementation, one would expect that a high degree of autonomy for local authorities coupled with a history of environmental concern and actions will provide a solid basis for local climate policy. One would expect this to depend not only on the degree of local autonomy in general but also on (1) the specific scope of autonomy with respect to local climate policy, and (2) the local capacity for action in terms of the resources and competencies of local authorities (compare Lafferty and Eckerberg, 1998).

In recent years several studies (Energie-Cités/Ademe Franche-Comté, 1996; Energie-Cités/Fedarene/Islenet, 1996; Climate Alliance, 1995; Collier, 1997) have been published comparing the situation in the field of energy policy and climate-policy in different European countries.

Energie-Cités, an association of European municipalities that promotes sustainable local energy policies, published a report (Energie-Cités/Ademe Franche-Comté, 1996) containing fact sheets on energy and municipalities in 15 European countries. The report identifies several important factors that explain the diversity between the countries investigated. First, the history and the resulting administrative organization of each country have determined the relative weight of local power. In each country the balance between national and local powers will determine the general room to manoeuvre enjoyed by municipalities, particularly in the energy sphere. Second, the presence, or otherwise the absence, of national energy resources has played a part in shaping energy systems and customs. And third, local climatic conditions have favoured intensive recourse to particular modes of energy: the production and distribution of heat is naturally more developed in countries where thermal needs are particularly acute. A fourth factor is the political circumstances, which have often had a strong impact on energy systems. Examples include the nationalization of gas and electricity at some point in the past in certain countries, the start of the liberalization of the energy market in other countries or the decision by countries to abandon nuclear energy.

The report concludes that a map of Europe illustrating the powers and jurisdiction of municipalities in energy matters would broadly distinguish two groups of countries:

- co untries in which the municipalities have considerable or very considerable powers: Sweden, Finland, Denmark, The Netherlands, Germany, Austria and (parts of) Italy. Urban energy planning and a truly municipal energy-management policy are often to be found in these countries;
- c ountries in which municipalities have few or limited powers: the United Kingdom, Ireland, France, Belgium, Spain, Portugal and Greece. Situations may vary within this group, but very few municipalities possess an urban energy planning system.

A project carried out at the European University Institute concentrated more specifically on climate change policies, including an examination of local action in five EU member states (Collier, 1997). This report concludes that climate change activities can be expected to depend on two main factors. First, the competencies in crucial areas of energy and transport, which affect the scope of actions by local authorities. Second, public sensibility both to environmental issues in general and climate change issues in particular. Collier suggests that local authorities are unlikely to get involved unless there is public and political support. On that basis, she expects more action from so-called 'leader' countries.

Together with a number of partners, the Climate Alliance (an association of concerned local authorities, see below) conducted a survey of municipal activities to reduce CO_2 emissions and of the national framework conditions for local action in four European countries (Austria, Germany, Italy and The Netherlands). The national framework conditions were analysed to identify obstacles to local climate policy. For the purposes of this chapter, the overall conclusions concerning the role of the national framework are particularly interesting. The report concludes that the influence of the national framework is highly relevant to the extent and intensity of municipal climate protection activities. However, in some cases there are severe legal obstacles to further activities, while in most cases the competencies of municipalities are not clearly defined or there is a lack of funding or know-how. Even under the existing framework conditions there are many options for municipal carbon-reduction strategies and measures that no municipality has actually completely exhausted. Nevertheless, local authorities could act much more effectively if the national and European framework conditions were set properly (Climate Alliance, 1995).

Collier (1997) reaches a similar conclusion on the importance of policy frameworks at the national and EU level. As 'unfavourable', she identifies budget constraints imposed on local authorities, the pursuit of low energy prices combined with a resistance to imposing carbon taxes, the absence of energy-efficiency standards and insufficient support for public transport.

To summarize the most relevant factors for our discussion here, the role of local authorities is determined by:

- the general autonomy of local authorities;
- their specific autonomy in crucial policy areas of energy and transport;
- public and political support for environmental and climate issues.

Appropriateness of the local level of governance in climate policy

Why is local government an appropriate level of governance for climate policy? In the introduction we gave three arguments for the importance of local climate policy. Two of the arguments given, that local authorities initiate local activities that result in greenhouse gas emissions and that they have responsibility in many areas of policy that are significant for climate policy, refer particularly to the criterion of effectiveness.

The EU captures the effectiveness criterion for the most appropriate level of governance for climate policy in the following formulation: *'to establish the level best suited to the type of pollution and the geographical zone to be protected'* (First ÉU Environmental Action Programme). This criterion refers to the extent that climate problems have their roots in local activities and local government's initiatives. Local policy therefore can be effective policy.

The argument that local authorities are the level of government closest to the citizen is in itself a normative criterion for local democracy. The normative criterion 'as close to the citizen as possible' is linked with effectiveness through public involvement and its consequences for 'raising awareness' and 'building support' for climate policy.

For instance, local authorities can deal very effectively with public involvement because *'as the level of governance closest to the people, they play a vital role in educating, mobilizing and responding to the public to promote sustainable development'*.

The question of why local government is an appropriate level of governance for climate policy brings us to the concept of subsidiarity. On the basis of the work in the SUSCOM project, Lafferty (2000) states that the principle of subsidiarity can be understood in two ways. One has its roots in a historical-philosophical discourse, where subsidiarity refers to the idea that superordinate decision-making bodies should only exercise powers that are ostensibly necessary for securing a 'better' (more effective, more efficient, more morally correct) state of affairs for individuals and institutions at a lower level of organization. The Maastricht Treaty states that: *'in areas which do not fall within its exclusive competence... the Community shall take action... only if and insofar as the objectives of the proposed action cannot be sufficiently achieved by the member states and can therefore, by reason of the scale or effects of the proposed action, be better achieved by the Community.'* The burden of proof in exercising this understanding of the principle is that the transfer of authority upwards gains its legitimacy from a necessary or desirable additional contribution that is not attainable through the efforts of lower-level domains.

Lafferty refers to the second connotation of subsidiarity that he distinguishes as more 'modern' and politically motivated. In his opinion, in many ways this interpretation turns the first connotation on its head. Instead of arguing why it is necessary and/or desirable to provide subsidiary authority at a higher level of governance, it is used to argue for the moral necessity of keeping decision-making authority as dose to local decision-makers as possible. What was originally a principle to justify a transfer of 'sovereignty' upwards, here becomes a principle to enhance democracy and protect sovereignty at a lower level: 'Decisions should be taken as close as possible to the citizens affected by the decisions.'

If we apply the principle of subsidiarity to the dilemma of 'decentralization versus the necessity of national direction of the transition' we have to be aware that the subsidiarity debate focuses on the division of powers between the supranational bodies of the European Union and the decision-making bodies of the member states. Secondly, the two interpretations of subsidiarity are interrelated. As Lafferty states, there is implicit reference to the more philosophical, legalistic connotation in the more political connotation. The decision that is to be taken closest to the people affected must be the best decision possible within the scope and functionality of that particular unit.

The potential changes in the role of local authorities in climate policies

Local policy-making is important for realizing a radical change in greenhouse gas emissions because local policies are important for the key aspects of transition processes that:

- de al with the use of space;
- de al with the built environment;
- re ly on household behaviour and consumption.

How important local policies will be in transition processes depends on how the 80 per cent greenhouse gas reduction goals are to be reached. In the two visions of the Dutch energy systems in 2050 (see Chapter 6), the role of government, and particularly of local government, differs quite appreciably.

In Vision A the role of government will be more reserved. Vision B assumes a stronger government with greater regulation. In the internationally-oriented 'global village' in Vision A, the role of local authorities will be smaller than in Vision B, which is more regionally oriented. The role of local spatial planning will be quite different in the two visions. The assumption in Vision A is that people live in large houses in spacious suburbs, while in Vision B people live in compact cities in smaller houses. In Vision B municipal land-use planning will be more restrictive. In Vision B renewable energy is strongly supported. The municipalities have to reserve space for the wind-power capacity that has to be installed on land. Furthermore, local authorities have to create the conditions for large-scale application of solar energy in the built-up environment.

As regards dwellings, local authorities bear responsibilities for energy efficiency, for instance through standards of insulation. In Vision B the local authority can make use of spatial measures that will favour energy-infrastructure measures such as district heating, compact building, passive solar heating and the integration of combined heat and power (CHP) generation. In Vision A a completely revised energy infrastructure will be needed.

In Vision B, public transport also plays a much larger role. Municipalities can play an important role in stimulating people to get out of their cars and into public transport. Finally, local authorities can play an important role in raising public awareness and creating support for reduction of greenhouse gas emissions. In Vision B, consumer behaviour is itself an essential component of reducing greenhouse gas emissions.

To conclude, Vision B envisages a more active role by local authorities in regulation, spatial planning, sustainable housing and sustainable production and consumption.

DUTCH MUNICIPALITIES AS AN EXAMPLE

The role of local authorities in Dutch climate policy

Because it is difficult to define a universal playing field, we will take a specific country as an example. Given the factors determining the role of local authorities in climate policy discussed in the previous section (general autonomy, specific autonomy and public and political support for the climate) we will briefly discuss the Dutch situation, applying the factors discussed above to the Dutch example.

General autonomy of the Dutch local authorities

Classifications of 'models' or types of local government systems (Hesse and Sharpe, 1991) place The Netherlands within the 'Napoleonic' system. This model is characterized by a history with a relatively high degree of centralized state control. Local authorities tended to be 'supervised' by mayors, who were in turn strongly associated with the central government. The Dutch constitutional system is generally labeled as a 'decentralized unitary state'. The unitary nature of this type of state is based on agreement between the three levels of government (national government, provinces and municipalities) and not on central government. Local authorities have freedom of initiative within the system of inter-administrative relations. They have the constitutional power to deal with matters of local concern as long as the local authorities take account of legislation passed by higher authorities. Many local authority activities are covered by legislation within a system of co-government.

Dutch local authorities have relatively limited tax revenues of their own and depend largely on central government for their resources. About half of this central funding takes the form of a specific transfer, or earmarked funding. The rest of the central funding is in the form of a Municipal Fund, which is an open-budget system with a budget ceiling, with its distribution depending on criteria such as the number of inhabitants.

Public and political support and specific autonomy

The Netherlands can be regarded as being among the 'leader' countries in environmental policy (Jänicke and Weidner, 1997; Organisation for Economic Cooperation and Development, 1995). Most Dutch municipalities have had an energy policy since the 1970s. In the early period the main aim of this energy policy was to conserve energy for financial reasons, given the rising energy prices. Energy saving was an aim in itself. Very little attention was given to the role that municipalities could play with respect to other energy users like businesses and citizens. From the end of the 1970s onwards there was a lot of discussion between national and local government about the way municipalities performed their local environmental tasks. The major problems were the lack of priority given to local environmental policy in general and the lack of local capacity in terms of resources and manpower. As a result of these discussions, during the 1990s municipalities received earmarked funding to bring their environmental policy to a basic and adequate level.

In the first National Environmental Policy Plan (NEPP) (1989) and the first white paper on energy saving (1990) municipalities were given a role as co-executor of energy policy. Energy policy was linked to sustainable development through key themes like the greenhouse effect, acid rain and the depletion of resources. To stimulate the implementation of the first NEPP, the ministry of the environment and the umbrella organization for local authorities (VNG) produced a framework for the implementation of the NEPP by municipalities (1990). This framework included basic mandatory and voluntary tasks in the field of energy, including the drafting of a municipal energy plan. NOVEM, an agency largely funded by the Dutch government which manages many national and international governmental programmes aimed at environmental improvement, increasing energy efficiency and introducing renewable energy sources, stimulated the drafting of these plans by subsidizing municipalities if they adopted a municipal energy-saving programme (known in Dutch as the GEA method). From 1st January, 1995 the first earmarked funding programme, the so-called BUGM funding, was followed by an extended programme, the VOGM funding, which ran until the end of 1997. In this latter programme, energy saving could be chosen as a voluntary priority task (known in Dutch as Ebb, which stands for energy-saving policy). Almost two-thirds of all municipalities joined this VOGM programme, which incorporated the GEA municipal energy-saving approach.

During 1995, however, the national government and the Climate Alliance identified a serious bottleneck; many municipalities were experiencing great difficulties in the implementation of the formulated policy due to obstacles in terms of finances, political commitment and organizational match.

At that time more than half of the municipalities had drawn up an energy-saving plan using the GEA method. This led to a new initiative, the so-called LOREEN programme, which was specifically designed to support municipalities and remove obstacles to the municipal implementation process. After the period of earmarked funding, many local governments had to improve their environmental policy and further extend their own responsibility.

When the new agreements were being concluded between national and local government, climate change was raised as an issue. All parties are now working on a so-called climate voluntary agreement. Many municipalities are interested in concluding an agreement on municipal climate policy with the national government in which municipalities will agree to give climate policy a prominent place on the local policy agenda in exchange for national government support (financial resources, knowledge, adequate legislation, etc.).

At the time of writing, about two-thirds of all municipalities have adopted the GEA approach. A survey shows that 70 per cent of these municipalities, for whom the planning period is coming to an end, have plans to develop follow-up policies (NIPO, 2000). In addition, nearly all municipalities that formulated policies are actually implementing them or are already in the phase of formulating new policies (NIPO, 2000). The willingness of Dutch municipalities to get involved in climate policy is also demonstrated by other initiatives. At the time of writing, one in six local authorities (around 100 in all) have joined the International Climate Union, an alliance of native populations in the Amazon basin and western local councils committed to the ambitious target of reducing CO_2 emissions by 50 per cent from 1987 levels by 2010.

The greenhouse gas reduction potential of Dutch municipalities

Municipalities have many options to reduce greenhouse gas emissions. The theoretical reduction potential of municipal policies depends on three factors; first of all, the extent to which a certain reduction option is within the area of competence of a municipality. This competence, as we have seen, will differ from country to country. Secondly, not all climate options within the competence of municipalities in a particular country are equally relevant to all local authorities. Some of the options are directly linked to certain characteristics of a local authority, such as having a new housing estate or an industrial estate or businesses with a certain level of

Table 10.1 Areas of local authority responsibility for various target groups

	Areas of local authority responsibility				
	Planning	*Construction and housing*	*Transport and traffic*	*Environment*	*Local government tasks*
1	2	3	4	5	6
Utility buildings	compact building, passive solar energy/zoning plan energy infrastructure, energy vision	Sustainable Construction and Energy Performance Standard (EPN)/ maintenance, information, covenants, etc.		energy saving and energy/environmental licence, environmental communication	energy saving, local sustainable government buildings/environmental care, subsidy conditions energy saving, public transport and traffic lights/public lighting plan
Households	compact building, passive solar energy/zoning plan energy infrastructure, energy vision	Sustainable Construction and Energy Performance Standard (EPN)/ maintenance, information, covenants, etc.		energy saving and sustainable energy/ environmental communication	
Traffic	decrease car mobility/ special planning, location policy/low traffic density neighbourhood/zoning plan		decrease car mobility/ traffic planning, parking	transport management at companies/ environmental policy cycle route plan	sustainable industrial estate improvement public licence transport transportation management own organization

(Contd.)

Table 10.1 continued...

1	2	3	4	5	6
Waste	residual heat utilisation/ structure plan, choice energy infrastructure			waste prevention and recycling at companies/ environmental licence sustainable industrial estate land fill gas extraction and kitchen and garden waste fermentation / waste disposal	
Energy sector	residual heat utilization/ structure plan space for sustainable energy/zoning plan	sustainable energy/ sustainable construction policy			
Industry	residual heat utilization/ structure plan, choice energy infrastructure			energy saving/ environmental licence sustainable industrial estate	
Agriculture	utilization CO_2 or residual heat/structure plan choice energy infrastructure space for sustainable energy/zoning plan			energy saving/ environmental licence	

energy consumption within its boundaries. Thirdly, there is the extent to which options are already used by municipalities.

In this section we will discuss the potential for greenhouse gas reduction by Dutch municipalities. A research project estimated the emissions of greenhouse gases related to activities that fall within the sphere of influence of Dutch local authorities to be more than 40 per cent of The Netherlands' total greenhouse gas emissions (Burger, Coenen and Menkveld, 2001).

The following summary outlines the most promising local options in fivel areas of local authority responsibility: town and country planning, construction and housing, traffic, environment, and local authority management tasks. The list of options given here is by no means exhaustive, but illustrates what can be attained and what is actually being achieved at the moment. In the following matrix the tasks are placed along one axis and the target groups of climate policy along the other. The various climate policy target groups identified in the Dutch Environmental Balance (state of the environment report) under the heading of climate change are utility building (or: Trade, Services and Government), households, transport and traffic, waste, the energy sector, industry and agriculture. The local authority playing field encompasses a mixture of options and available instruments under direct municipal influence and indirectly via target groups on the one hand, and its areas of responsibility on the other. Table 10.1 presents a summary of important local options for greenhouse gas reduction, given the possibilities that the local government has to influence them, in five areas of responsibility. We will briefly discuss these options, their reduction potential and how this potential has been exploited in practice up to now.

Town and country planning

Spatial planning can extend the possibilities for waste heat utilization by locating consumers of heat (for instance, a residential area) dose to a heat producer (for instance, a power plant or a waste incinerator). This means that in the construction of the energy infrastructure in new or redeveloped residential locations, choices have to be made regarding the heat-distribution network. The Dutch national government expects a substantial contribution to CO_2 reduction from the use of residual warmth and the application of combined heat and power (CHP). The point of departure for the national government is that installed CHP capacity should double between 2000 and 2010 from 8000 MW to 15,000 MW, which could represent a total CO_2 reduction of 5–10 Mtonne CO_2 (Uitvoeringsnota Klimaatbeleid, 1999).

At the moment, the availability of waste heat is scarcely considered in decisions on the location of potential consumers of that heat, such as new residential areas, industrial areas or agricultural greenhouses. The subject is mentioned in national and provincial spatial planning documents, such as the Fifth National Spatial Planning Framework, but in practice not much happens yet.

Van der Waals *et al.* (2000a) investigated the decision-making processes surrounding the design of 26 larger new residential areas, the so-called VINEX areas, with more than 4,000 houses. In 15 of the 26 locations, a 'heat network' was either chosen or a strong preference was expressed for such a network. The possibilities in smaller new housing areas are expected to be less positive. The areas covered by the study represent 226,000 houses already built or to be built between 1995 and 2005. The total number of new houses built in The Netherlands each year is about 100,000, so the areas studied represent about a quarter of the total number of new or planned houses. Urban planning instruments like compact building and passive utilization of the sun (orientation of plots to the sun) can also contribute to CO_2 reduction. The same study shows that in many of the VINEX locations more than 50 per cent of the plots are oriented towards the south.

Municipalities play an important role in creating space for wind energy. Wind energy with a primary energy-saving potential of about 50 to 100 PJ (CO_2 reduction of about 3 to 6 Mtonne) could become one of the most important sustainable-energy options in the longer term. Unfortunately, few Dutch municipalities have an active policy towards wind energy. So few, in fact, that the national government has threatened to force municipalities to install wind turbines.

In theory, urban planning could help to reduce car mobility. The impact of these measures is uncertain. People seem to be less sensitive to distance. The potential reduction of car mobility in any urban area is limited since it doesn't depend on the number of car movements but on the number of kilometers traveled by car. Only about 6 per cent of all kilometers traveled in a car involve journeys over a distance of 5 kilometers or less, journeys that could be made by bicycle. However, a housing area should offer the organizational infrastructure to make car-sharing, car-pooling and teleworking possible. The research referred to above shows that in all the VINEX locations studied, alternatives to using the car have been stimulated, but only in a few cases alternatives to car use or car ownership.

Construction and housing

The reduction potential of sustainable construction is large; 40 to 50 per cent of emissions by households and utility buildings are caused by the use of energy for heating. So if sustainable construction measures can reduce the energy use in buildings by 10 to 15 per cent, we are talking about a serious reduction potential of several Mtonne CO_2 per year.

In 1998, 32 per cent of all newly built houses complied with standards that exceeded existing legal sustainable-building requirements laid down in a sustainable construction performance standard. The extra measures involved the application of all cost-neutral or cost-saving measures set out in the National Package Sustainable Building and all variable measures involving extra investment of up to 2,000 guilders (Beleidsprogramma 2000–04).

In the existing housing stock, there were example projects for 56,000 houses in the period 1995–98 and subsidies were promised for sustainable renovation. An inventory of energy saving in housing renovation plans covering about 70 projects, which involved around 1,000 houses, shows that about one-third of the plans included energy saving as a separate goal. Some smaller sustainable measures relating to buildings, like solar energy, solar boilers and heat pumps, are only used on a small scale or in experiments (Van der Waals *et al.*, 2000b). All in all, we can conclude that measures for sustainable construction are frequently used, especially in one-third of new-built houses, but that it is certainly not common practice.

Transport and traffic

Options in traffic range from decreasing car mobility (encouraging cycling, car sharing, teleworking, transportation management) to encouraging public transport and car pooling to electric vehicles and biofuels. In the area of transport and traffic, traffic planning, spatial organization and parking policy are used to reduce car mobility. Other options are the insertion of transport management in environmental licences of businesses or in sustainable industrial estates.

Stimulating public transport is probably the most promising option in terms of reduction potential in this policy area. It is unclear how much influence municipalities have here. Municipalities can also stimulate car sharing, car pooling and Park & Ride options by creating the infrastructure.

Much of what is happening in this policy field is targeted at parking. However, the motivation behind parking policy is seldom climate policy but more to do with tackling congestion in inner cities. Municipal measures are concerned with improving the logistics of urban distribution.

Environmental policy

Municipalities can prescribe energy measures in the environmental permits of small businesses and the service industry. National government is responsible for arrangements with the energy-intensive industries through a voluntary agreement, the so-called Convenant Benchmarking, which is expected to cover about 80 per cent of all industrial energy use. However, industry in The Netherlands uses so much energy (1,100 PJ) that the remaining 20 per cent still represents a large part of the total energy use in The Netherlands and accounts for about 5–10 per cent of all greenhouse gas emissions. Furthermore, the service sector is also responsible for about 13 per cent of all greenhouse gas emissions in The Netherlands. According to the national government, energy saving has not received enough attention in environmental permits issued by municipalities in recent years. Energy-saving measures in permits go beyond just applying energy measures in the production processes. They also relate to indirect energy savings by reducing mobility, the economical use of resources and waste prevention.

Dealing with traffic management in permits is difficult in practice, although the reduction potential is large. One-third of all kilometers traveled by car involves commuter traffic. Traffic management could reduce this number by between 6 per cent and 20 per cent. Traffic management can also be applied to the transport of goods by firms.

Waste prevention has been a topic of discussion in The Netherlands for many years and a great deal of experience has been gained. An important driver for waste prevention in (larger) businesses has been the 'pollution prevention pays' principle. Environmental permits do not devote much attention to waste prevention and resource use although municipalities do have the legal power to include them. There is a reduction potential here for the stragglers and companies without environmental care systems.

Separate waste collection is a crucial condition for recycling. In 1995, 42 per cent of domestic waste was collected in separated streams and the goal is 60 per cent by 2000. Especially the recycling of products from energy-intensive industry (like metals, paper and glass) contributes to energy savings in production, while recycling of synthetic materials helps to curb the use of oil products and CO_2 emissions.

Methane emissions from landfills still contributed 4 per cent of the total volume of greenhouse gas emissions in The Netherlands in 1998, despite the use of landfill gas extraction. The use of landfill gas has a large reduction potential which could be used more, but is decreasing because of the separate collection of kitchen and garden waste which removes organic waste from the waste being dumped.

Local authorities also play an important role in communication about the environment. Climate change could be an aspect of environmental communication in raising awareness that climate policy and energy saving is necessary and that the target group has a crucial role to play in it. Local authorities can also inform climate policy target groups about laws, subsidies, etc. Finally, the municipality could communicate the example it sets in the field of energy saving.

Municipal management

A municipality can take measures in municipal buildings (swimming pools, sports facilities, theatres, offices, etc.). The reduction potential is limited. It is estimated that only 6 per cent of the energy use and emissions of public buildings can be attributed to municipal buildings.

There are many options for energy saving in public lighting (street lights, traffic lights). Public lighting uses 1,000 million kWh per year, and three-quarters of this is the responsibility of municipalities. It is believed that about 300 million kWh (1.5 Mtonne CO_2) could be saved at the municipal level. It is difficult to get an overall picture of how many energy-saving measures are already applied by municipalities. Many municipalities seem to have experience with one technical option and not others, but here also options don't seem to be commonly applied.

Lessons to be learned from the Dutch example

The use of the playing field

One could ask whether municipalities in a country are exhausting their climate options. A Dutch research project postulated the assumption that local government in The Netherlands could play a greater role than it currently does in reducing the country's greenhouse gas emissions. Many local climate options are known in theory. Unfortunately, evaluation of the climate options identified in this chapter shows that many options for climate reduction are not availed of even though municipalities have the competence to employ them. Actual practice turns out to be rather fractious.

The evaluation research shows that many options for greenhouse gas reduction do not become common practice (Burgers *et al.*, 2001b). An explanation lies in the obstacles faced by local authorities that want to implement these options:

- political barriers; a lack of political support for a particular policy option;
- communicative barriers; sometimes the policy direction or ambitions are unclear;

- organizational or institutional barriers; such as the lack of co-operation between different departments in the same local authority or dependence of local authorities on co-operation with other actors;
- resource barriers, such as manpower and finances, and the low expectations of municipalities for certain options.

Since 1996, the LOREEN programme has provided considerable support for Dutch municipalities that applied for it in overcoming these barriers to local energy and climate policy. Evaluation of this stimulation programme shows that development and implementation of both energy and climate policy unproved considerably in these municipalities (Coenen *et al.*, 2001). However, what is evident from the brief review of the climate options is that many of the options for local climate policy in fact fall outside the area of environmental tasks. This touches on the dilemma of 'transition through co-benefits' *versus* 'inevitability of a special climate change approach' and the dilemma of whether to separate climate change from sustainable development policies.

Furthermore, the possibilities for a specific target group to reduce greenhouse gas emissions always rum out to be divided over a number of areas of responsibility, which leads to the conclusion that local climate policy is pre-eminently an integrated policy. If such an integrated policy is to be implemented, then institutional barriers within the local authority will have to be overcome and local climate policy will have to be approached more as an integrated problem extending also to non-environmental policy areas.

Potential for change

In the introduction we referred to two options for a trend break in the contribution of local authorities to climate policy. The first option is to enlarge the role of local authorities by expanding their competencies and playing field. The second option is to optimize the use of the municipal playing field. These options accord with the results of international research.

As we saw in the second section of this chapter, many of the cross-country comparative research projects look for factors in national policies that hinder local climate policies. Clearly these factors exist. In The Netherlands, the role of the local authorities in Dutch climate policy and climate research has hitherto remained somewhat under-illuminated. For instance, little attention is paid to the local level in the national climate strategy as set out in the Climate Policy Implementation White Paper (1999).

Furthermore, the playing field is not static and the possibilities for local climate policy can change due to developments in society at large. The liberalization of the energy sector, the trend towards more decentralization and local standard-setting in environmental and energy policy and the changes in local democracy are particularly important for local climate policy in The Netherlands. The overall conclusion from research (Burger *et al.*, 2000) that investigated how these developments may change the municipal playing field was that through the influence of societal changes the role of local authorities will become more important. Public expectations are higher and through processes of liberalization of the energy markets the co-operation with the profit-based energy firms will change. Processes of decentralization will mean that local authorities can rely less on national government. All these developments point to a greater role for local authorities in climate policy in the near future.

Trend breaks and dilemmas

In this chapter we considered two options for a trend break in the contribution of local authorities to climate policy: (1) enlarge the role of local authorities by expanding their competence and playing field, and (2) optimize the use of the municipal playing field.

We feel that such a trend break is essential for a transition towards a climate-neutral society. An overall reduction of emissions by 80 per cent can never be achieved without a prominent role for the local level of governance. The Dutch example shows that 40 per cent of total Dutch greenhouse gas emissions are related to activities that fall within the Dutch local authorities' sphere of influence. This percentage is probably lower than in other countries, where the general autonomy of municipalities, and their specific autonomy in the fields of energy and climate policy, are larger.

The first trend break option is to enlarge the role of local authorities by expanding their competence and playing field. With respect to the specific dilemma of decentralization versus the need for national direction of the transition introduced in Chapter 7, it is inherently doubtful whether local government is an appropriate level of government for climate policy.

In the second section of this chapter we said that arguments in favour of a greater role for local authorities in climate policy could be based on both legitimacy and effectiveness criteria:

- the legitimacy criterion is that climate change policies should be developed and implemented as close to the citizens as possible;
- the effectiveness argument is that a large part, 40 per cent in the case of The Netherlands, of total greenhouse gas emissions are related to activities that fall within the sphere of influence of Dutch local authorities. This illustrates the fact that many climate problems have their roots in local activities and local governments' initiatives;
- this influence could be increasing given developments in society at large. In particular liberalization of the energy sector, the trend towards further decentralization and the changes in local democracy will cause the role of local authorities to change in the coming years. The role of local authorities will become more important and local authorities will therefore have to be more active.

As shown, a lot of attention is given in international literature to the national factors that may hinder local climate policy. The suggestion is that by removing these obstacles local authorities could act much more effectively if the national and European framework conditions are set properly, However, there are some serious dilemmas in striving towards a larger role for local authorities. Any form of decentralization of climate policy or giving local authorities a larger responsibility poses problems. Climate change is a difficult political issue at the local level because of two dilemmas mentioned in Chapter 7. The long-term perspective needed in climate-change policies is difficult for politics, which focuses more on urgent short-term local priorities. The relatively short-term character of local politics and the focus on urgent local priorities make it difficult to separate climate change from sustainable development policies. Furthermore, these characteristics of local politics not only make it difficult to achieve radical change, but also make it also difficult to prepare for radical changes in 30–40 years.

The second option, optimizing the use of the existing playing field, is specific to country and local authority. We discussed this trend break on the basis of the Dutch example. We sketched the role of local climate policy in The Netherlands, the local climate options of Dutch municipalities, their total potential for reducing greenhouse gas emissions and their use of the playing field.

This option would involve optimizing the contribution of local authorities within the existing national framework, presuming a large willingness to act among the majority of municipalities. A lack of competence should not be confused with an unwillingness to act. We believe that a greater contribution by local authorities should not be seen solely from the perspective of hindrances from outside the organization or of a lack of willingness to act, although both problems exist, but that of an optimal use of local climate policy options.

To enhance the contribution of local governments to climate policy, a systematic integrative approach is necessary. Evaluation of the experiences with local climate options and of the growing significance of local climate policy suggests the need for a larger contribution by local authorities. Improvements should be expected not only from climate options themselves but also from an organizational and process perspective. The local level is the pre-eminent level where policy decisions find concrete implementation and thus have a direct impact on daily practice. When policy decisions are being made it must rapidly be made dear what the consequences are for the emissions of greenhouse gases. This must make an integrated climate policy possible, piggy-backing with many other policy areas. Such an approach needs several basic steps or elements, which build on basic steps from total quality management and environmental care systems:

1. a policy document that articulates the political commitment to local climate policy;
2. plans and programmes to implement climate policy inside and outside the organization;
3. integration of these plans in daily policy practice and in the organizational culture;
4. monitoring the performance of local climate policy;

5. offering education and training to raise the awareness and understanding of the climate problem;
6. publishing information on local climate policy performance.

Following these basic steps would do justice to the potential options in climate policy and integration in other policy areas. Research (Menkveld *et al.*, 2001) shows that the third step is the most crucial one in introducing 'climate care' throughout all tiers of the local authority. However, political commitment and drawing up plans are preconditions for integrating these plans in daily policy practice.

To answer the question of what local authorities can and should do to contribute to a 80 per cent greenhouse gas reduction, we suggested two pathways. The first path is enlarging the role of local authorities by expanding their competence and playing field. The second path is using the existing municipal playing field as advantageously as possible. Expanding the role of local authorities would come down to asking many local authorities to do their bit for problem-solving. The observation that many municipalities only pursue some of all the possible climate policy options addresses the dilemma of a 'small numbers of decision-makers' trajectory versus a 'decentralized multiple and diffuse decisions' trajectory. One strength of a larger role for local authorities could be wider public support because of the smaller distance between local politics and local people.

Local climate policies encompass a broad range of policy fields. Expanding the role of local authorities in climate policy touches on the dilemma of 'transition through co-benefits' *versus* 'inevitability of a special climate change approach.' The relatively short-term character of local politics and the focus on urgent local priorities make it difficult to separate climate change from sustainable development policies. These characteristics of local politics also make it difficult not only to achieve radical change in the short term but also to prepare for radical change in 30–40 years.

❑❑❑

11

The Future

MUCH OF THIS BOOK HAS in one sense been historical, attempting to outline the various changes that have been wrought by humans on the face of the earth over the last few million years as levels of population and the nature of technological progress have changed. Nonetheless, from time-to-time reference has been made to the future and to the uncertainties that surround it. Given the rapid expansion in human numbers and the increasing sophistication and pervasiveness of certain technological developments, the pace of change over the next decades is likely to become such that the human impact will be greatly magnified.

There is now an unprecedented interest in the environmental future, and few people are unaware of the threat and potential dangers posed by nuclear pollution, species extinction, acid rain, the carbon-dioxide problem and desertification. It is for this reason that various attempts have been made to formulate informed predictions of likely changes as a guide to governments and administrators. One of the most notable of these is the report prepared and first published in 1980 as *Global 2000*, an American compilation for the President of the USA (Council on Environmental Quality, 1982). Some of its findings on such matters as the extinctions of species that may result from deforestation in the moist tropics have already been referred to. The report is based on an enormous amount of informed research by a large number of often distinguished scientists. Although it is careful not to be excessively negative, and weighs up various alternative scenarios, its basic conclusions are fundamentally pessimistic:

> If present trends continue, the world in 2000 will be more crowded, more polluted, less stable ecologically, and more vulnerable to disruption than the world we live in now. Serious stresses involving population, resources and environment are clearly visible ahead. Despite greater material output, the world's people will be poorer in many ways than they are today. For hundreds of millions of the desperately poor, the outlook for food and other necessities of life will be no better. For many it will be worse. Barring revolutionary advances in technology, life for most people on earth will be more precarious in 2000 than it is now—unless the nations of the world act decisively to alter current trends.

In 1984, however, Simon and Kahn edited a volume entitled *The resourceful earth — a response to global 2000*, which took a very much more optimistic view, and which included contributions from a variety of distinguished scientists, including some eminent geographers. Note the alternative view of the future which they propose:

> If present trends continue, the world in 2000 will be *less crowded* (though more populated), *less polluted, more stable ecologically* and *less vulnerable to resource-supply disruption* than the world we live in now. Stresses involving population, resources, and environment *will be less in the future than now*.... The world's people will be *richer* in most ways than they are today.... The outlook for food and other necessities of life will be *better*... life for most people on earth will be *less precarious* economically than it is now.

There are many possible reasons why there should be such a wide disparity in attitudes to the likely future of the earth. Amongst these are a paucity of background data on the present and past state of the earth, the problems of extrapolating data into the future by curve-fitting procedures, the crudeness and assumptions of so many models, the complexity of interactions referred to in the last section, uncertainty about the nature of technological developments and possible political constraints on the use of fossil fuels and nuclear power. Inevitably also differences in opinion may result from ideological and political differences between authors. However, whatever the standpoint of the investigator, it is unquestionable that our fundamental knowledge is still far too slim on which to build reliable prognoses of the future, and that geographers and practitioners in allied disciplines need to seek still more information.

The climatic future

Such is the uncertainty surrounding the climatic future that the *Global 2000* (Council on Environmental Quality, 1982) report considered three different scenarios feasible:

Case 1 (No change). Yearly rainfall and temperature statistics would be similar to those of the 1941–70 periods, with occasional droughts in areas like the Sahel, India and the USA. This scenario might result if the warming trends induced by the build up of CO_2 were cc unteracted by natural cooling tendencies.

Case 2 (Warming). Global temperatures would increase by 1°C, principally as a result of a warming trend caused by elevated CO_2 levels being sufficiently large to outweigh the effects of any natural cooling. Most of the warming would take place in polar regions and the higher middle latitudes, with only slight warming in the tropics. Annual precipitation would increase by 5–10 per cent on a global basis, hut there would be more likelihood of US drought conditions similar to those of the Dust Bowl years of the 1930s.

Case 3 (Cooling). As a result of changes in solar and volcanic activity, global temperatures would decrease by 0.5°C in the higher and middle latitudes. In general precipitation amounts would decline, and variability increase. Stormtracks, and the precipitation they bring, would shift towards the equator, improving conditions in the upper latitudes of the great deserts and worsening them on their equatorial sides. Severe monsoon failures would be more frequent in India, with severe droughts in the Sahel.

Assessment of the future scale and trend of climatic change is thus full of imponderables. This has been well stated by Warrick (1988):

> Prediction of the rate at which atmospheric CO_2 will accumulate in the atmosphere depends largely on assumptions regarding future fossil fuel consumption and on knowledge of the global carbon cycle. Added to CO_2 are changes in the atmospheric concentration of other 'greenhouse' gases—methane, chlorofluorocarbons, nitrous oxide, ozone—which in aggregate, are becoming just as important relatively speaking, as CO_2. Taking into account the projected concentrations of these gases (with the associated

> uncertainties), there could possibly be an 'equivalent' doubling of CO_2 by as early as the 2030s.... But changes in climate will not follow stepbystep. Owing to the thermal inertia of the oceans, changes in global surface temperature will probably lag behind by several decades. Thus the question of 'when' is clouded.
>
> The question of 'where' is even more problematic. Climatic changes will vary in direction and magnitude from one agricultural region to the next. Unfortunately, GCMs (General Circulation Models) cannot yet provide reliable predictions of climatic change at regional scales....

Future energy scenarios are notoriously variable, and hence there are huge uncertainties with respect to fossil-fuel consumption. Major problems include the prediction of economic growth in different economies and the indeterminate nature of future energy choice (as, for example, between nuclear and conventional power generation) (Bach, 1988).

In addition, there are major limitations in our ability to predict what will happen to the CO_2 released by the burning of fossil fuels and by deforestation and soil humus destruction. Major uncertainties surround the role of oceans as sinks for CO_2, the role of oceanic biota in regulating global carbon dynamics (see Clark *et al.*, 1982), and the role of different terrestrial biomes in extracting CO_2 from the atmosphere.

A major problem lies in the reliability of GCMs, many of which have been over simple and have many debatable assumptions. Dickinson (1986) has urged caution:

> The GCMs have tended to use submodels that are considerably oversimplified.... Oversimplified ocean and sea-ice models, together with questionable descriptions for ice and snow albedos, suggest that ice-albedo radiative feedbacks in GCMs are of doubtful accuracy. Likewise, the models used for land surface water, snow budgets, and evapotranspiration have been too simple to allow quantitative evaluation of the effects of realistic changes in land surface processes on climate.

It is for these sorts of reasons that GCM scenarios for the future of climate have been so variable (Schlesinger and Mitchell, 1985).

The real world is more complex than the models that are designed to represent it, and the ocean-land-cryosphere-biosphere is governed by numerous feedback mechanisms which respond to changes in atmospheric composition and temperature. Macdonald (1982) lists the following for consideration:

1. *Temperature-infrared radiation feedback (negative)*
   As the temperature rises so does the outgoing infrared radiation.
2. *Water-greenhouse feedback (positive)*
   As temperature rises, the amount of water vapour increases since relative humidity is dependent on temperature. An increase in water vapour content increases the infrared blanket of the atmosphere and this causes a further temperature increase.
3. *Ice and snow cover albedo feedback (positive)*
   As temperature rises snow cover and ice will melt, decreasing the earth's surface albedo and thereby increasing the absorption of sunlight.
4. *Cloud cover feedback (positive)*
   As temperature rises, cloud cover becomes greater. This traps infrared radiation and so the temperature rises.
5. *Cloud cover feedback (negative)*
   As temperature rises producing more cloud, the atmospheric albedo increases, and this tends to reflect incoming solar radiation and so to lower the temperature (see Idso, 1982).
6. *Carbon-dioxide biofeedback (negative)*
   Increasing the carbon dioxide concentration in the atmosphere will increase the growth of the biosphere, which will fix carbon via photosynthesis and so tend to reduce the atmospheric level.
7. *Iceberg formation feedback (negative and positive)*
   As temperature rises, ice sheets will break up, producing, at least initially, large numbers of icebergs which will increase the earth's albedo until all the ice melts. Eventual

disappearance of the ice sheets and icebergs will lower the earth's surface albedo which will increase insolation absorption and increase surface temperature.

8. *Temperature-convection feedback (negative)*

   As temperature is increased at the ground, convection will increase, carrying sensible heat away from the ground and cooling it.

9. *Evaporation-precipitation feedback (negative and positive)*

   As temperature increases the intensity of the hydrological cycle will increase. Increased evaporation will carry off the additional temperature rise in latent heat, while increased precipitation will be working to return that same energy.

10. *Sea ice—ocean current feedback (positive)*

    The bottom-water producing currents are driven in the Antarctic by the formation of sea ice in the Weddell Sea. An increase in temperature will inhibit or stop sea-ice formation and remove the pump which drives these currents. As a result the cold bottom waters will not be replenished and an increase in ocean temperature will occur, leading to a loss of CO_2 from the oceans and a further increase in temperature.

11. *Biomass-albedo feedback (positive)*

    Stimulation of new growth as a response to increased temperatures will cause a decrease in albedo which will stimulate a temperature rise.

A potential problem in forecasting future climates is that change may take place not solely in a gradual and progressive manner, but by stepwise changes across thresholds. Positive feedbacks may be involved. A notable example of this type of possibility is Flohn's (1982) contention that a global warming of 4°C will cause a sudden warming of central Arctic sea ice. Because, in his view, the Antarctic ice will be less radically and less quickly affected, a globally asymmetric cryosphere will result, which will thereby cause a further profound change in the global general atmospheric circulation. Similarly, Bell (1982) has suggested that a marked warming in high latitudes could trigger the release of large quantities of methane, currently trapped in permafrost, subsea permafrost, and sea ice, thereby adding significantly to the greenhouse effect (see also Nisbet, 1990).

A comparable position has been taken by Woodwell (1990: 5), who believes:

> The speed of the warming may be determined less by human activities, unless rapid steps are taken to stop it, than by the warming itself, the climatically caused destruction of forests, and the oxidation of organic matter in soils and bogs. As the warming progresses and trees the at the warmer and drier margins of their distributions, organic matter from plants and soils will be released as carbon into the atmosphere, speeding the warming.

A second method of climate prediction is the climate analogue scenario, which is based on the use of a suitably defined ensemble of warm years from the recent instrumental record, which is then compared with either the long-term mean or a similarly defined cold-year ensemble (Webb and Wigley, 1985). This was the approach adopted by Wigley *et al.* (1980), who sought to predict the pattern of future temperature and precipitation changes by comparing the five warmest years in the period 1925–74 with the five coldest. As they recognized, however, they were dealing with a temperature shift of only 0.6°C, so the patterns obtained would not necessarily be representative of an even warmer climatic future.

Furthermore, as Webb and Wigley (1985: 242) point out, 'The use of past warm periods as potential analogs of a world with high CO_2 concentrations is based on the assumption that, given similar boundary conditions (as represented by the state of the oceans, the land surface and the cryosphere), the general circulation of the lower atmosphere will respond in a similar way, even with differing forcing mechanisms.' Such an assumption may not he valid.

A third approach to climatic prediction is the study of selected past warm periods. Flohn and Dansgard (1984), for example, examined the Early Medieval optimum, the Holocene optimum (at around 6000 years ago), the Last Interglacial (*c.*120 000 years ago), and the Late Tertiary. This method holds considerable promise, though there are major limitations imposed by data gaps and dating uncertainties for many parts of the world. Furthermore, boundary conditions may have been different in the Pliocene of the Late Tertiary, largely because the effect that the Himalayan uplift has had since that time on the location of the jet streams. Even the status of the Holocene 'altithermal' or 'hypsithermal' can be questioned, and Webb and Wigley (1985: 252) believe that:

> More data (and derived temperature estimates) are needed before it can be demonstrated that the global mean temperature was higher at 6000 B.P. than it is today. The current data suggest that the global mean temperature at 6000 B.P. was probably within 1°C of today's temperature.

Schneider (1984: 190) has also expressed doubts about using the altithermal as an analogue, stating that:

> Unless we can be sure that the Altithermal warm period was caused by an increase in CO_2 (or some equivalent annual, global scale external forcing) we cannot argue convincingly that a global warming induced by burning fossil fuels would result in similar regional climatic shifts—assuming, of course, we could know reliably what Altithermal seasonal and regional climatic patterns actually were.

The spatial patterning of change

Most models indicate that the increase in global surface temperature for a doubled CO_2 level will be around 1.5 to 4.5°C (Houghton *et al.*, 1990).

The various GCMs indicate that warming will not occur in a regular and standard manner all over the world. There will be major differences between different regions. Models agree that there will be enhanced warming in higher latitudes in late autumn and winter. For example, the United Kingdom Meteorological Office GCM suggests that in the northern hemisphere winter there will be a great swathe of land across from the north-eastern parts of North America to Central Europe, the Arctic Ocean and Siberia, where warming may amount to as much as 6–12°C. Likewise the south polar regions show a comparable pattern in the southern hemisphere winter (June, July, August). In the same months parts of the western USA and Central Asia will also show pronounced enhancements in their temperatures. The various GCMs also all tend to suggest that warming in the tropics will be smaller than the global mean and will vary little with season, being typically 2–3°C.

Simulations of the pattern of precipitation change are rather more complex and show marked differences between different GCMs. However, more recent models 'produce enhanced precipitation in high latitudes and the tropics throughout the year, and in mid-latitudes in winter' (Houghton *et al.*, 1990: 145). Changes in the dry tropics are generally small, with both increases and decreases, though most models simulate an enhancement of precipitation associated with a strengthening of the south-west monsoon in Asia. The pattern predicted by one GCM, the United Kingdom Meteorological Office model.

Predictions of precipitation change based on warm analogue years and from the reconstruction of Holocene anti-thermal conditions show less complex patterns. However, both methods tend to suggest that global warming will cause drier conditions in the huge agricultural region of the Central United States of America.

One of the few reasonably comprehensive attempts to predict future precipitation changes by combining the evidence from palaeoclimates, modern analogue data and GCMs is provided by Budyko and Izrael (1991). They argue on the basis of palaeoclimatic data that under conditions of warmth comparable to those that existed in the Pliocene precipitation would increase (by up to 30 cm per year) over most of Eurasia and the Sahara. On the other hand, under conditions comparable to those of the Holocene Atlantic Optimum, with temperatures up to l°C warmer than the present, there is a large zone (between 50 and 30°N) where precipitation levels would decline (by up to 20 cm per year in Central North America). There would, however, he improved moisture conditions in the subtropical regions (between 10 and 20°N) and in higher latitudes (more than 60°N). This pattern of changes is attributed to a northward shift both of the ITCZ and of westerly cyclone tracks. In sum, Budyko and Izrael believe that under conditions of marked warming both the mid-latitude and lower-latitude arid zones of the northern hemisphere will be wetter, whereas under conditions of less marked warming (*i.e.,* by around 1°C) areas like the Sahara and Thar will become moister, but areas like the High Plains of the USA or the steppes of the CIS will become drier.

Further consequences of a global warming

Tropical cyclones (hurricanes) are highly important geomorphological agents, in addition to being notable natural hazards, and are closely related in their places of origin to sea temperature conditions. They only develop where sea-surface temperatures are in excess of 26.5°C. Moreover, their frequency over the last century has changed in response to changes in temperature, and there is even evidence that their frequency was reduced during the Little Ice Age (see Spencer and Douglas, 1985). It is, therefore, likely that as the oceans warm up, so the geographical spread and frequency of hurricanes will increase. Furthermore, it is also likely that the intensity of these storms will be magnified. Emanuel (1987) has used a GCM which predicts that with a doubling of present atmospheric concentrations of CO_2 there will be an increase of 40–50 per cent in the destructive potential of hurricanes.

An increase in hurricane intensity and frequency will have numerous geomorphological consequences in low latitudes, including accentuated river flooding and coastal surges, severe coast erosion, accelerated land erosion and siltation, and the killing of corals (because of freshwater and siltation effects) (De Sylva, 1986).

However, it needs to be pointed out that the Intergovernmental Panel on Climate Change (Houghton *et al.*, 1990) was somewhat equivocal on the question of the extent to which warming would stimulate cyclone activity, stating:

> Although the area of sea having temperatures over this critical value [26.5°C] will increase as the globe warms, the critical temperature itself may increase in a warmer world. Although the theoretical maximum intensity is expected to increase with temperature, climate models give no consistent indication whether tropical storms will increase or decrease in frequency or intensity as climate changes; neither is there any evidence that this has occurred over the past few decades.

Of considerable importance in terms of both climatic feedbacks and the nature of future sea level changes is the response of the Arctic and Antarctic ice masses to a global warming.

With respect to the Arctic ice, the situation is controversial and unclear. On the one hand, there is the view of Budyko (1982: 233) who believes that 'If the mean global temperature rises to 2–3° higher than in pre-industrial time, it will be enough for the many-year-ice in the Arctic to disappear.' By contrast, Flohn (1982) believes a temperature increase of 4–5°C would cause an ice-free Arctic Ocean, creating conditions broadly analogous to those of the Pliocene. These views are disputed by Parkinson and Kellogg (1979), who contend that while a 5°C atmosphere temperature rise would cause Arctic pack ice to disappear in August and September, it would reappear during the winter months. Bentley (1984) argues that in all probability no less than a 10°C temperature increase would be required to remove Arctic sea ice in summer.

Equal uncertainties surround the future of the West Antarctic ice sheet. One extreme view has been postulated by Mercer (1978), who believes that predicted increases in temperature at 80°C would start a 'catastrophic' deglaciation of the area, leading to a sudden 5 m rise in sea level. A less extreme view has been put forward by Thomas *et al.* (1979). They recognize that higher temperatures will weaken the ice sheets by thinning them, enhancing lines of weakness, and promoting calving, but they contend that deglaciation would be rapid rather than catastrophic, the whole process taking 400 years or so. Robin (1986) takes a broadly intermediate position. He contends that:

> A catastrophic collapse of the West Antarctic ice sheet is not imminent, but better oceanographic knowledge is required before we can assess whether a global temperature rise of 3.5°C might start such a collapse by the end of the next century. Even then it is likely to take at least 200 years to raise sea level by another 5 metres.

Predictions of the rate of ice sheet response to global warming are highly problematic because they involve the use of models with many assumptions or unknowns. Many factors are involved, including the rate at which accumulation occurs. This is crucial in the context of Antarctica, where warmer temperatures could cause more snow deposition in an environment which is currently extremely arid. Such increased accumulation might offset increased rates of ablation. Budd (1991) has calculated that increases of around 4°C in temperatures in the Antarctic would lead to a 30 per cent increase in net accumulation. Also important, however, are varying snow surface albedos, variability in rate of calving of outlet glaciers, the protective and buttressing role of ice shelves, and the geometry and slope of the ice sheet (Huybrechts *et al.*, 1990). The rapidity with which ice sheets and glaciers decay (and hence sea level may rise) depends greatly on the terminal environment and certain topographic thresholds (Sugden, 1991).

Alpine glaciers are highly responsive to climatic change, as is made evident by their frequent and rapid fluctuations during the course of the Holocene (sec Grove, 1988). Substantial neoglacial advances and retreats have been caused by quite limited fluctuations in temperature and/or precipitation. The historical picture of glacial response to such fluctuations is, however, complex, and this behoves us to exercise caution in predicting the future picture of response to just one control of glacial state, *i.e.,* warming. Changes in precipitation (which control rate of accumulation) and cloudiness (which can control rate of ablation) will also be significant factors in determining glacier equilibrium, while topographic controls, the effects of size and varying propensity to surging will create local complexities of response.

Bearing such caveats in mind it is nonetheless highly likely that most alpine glaciers will show increasing rates of retreat in a warmer world, and given the rates of retreat experienced in many areas in response to the warming episode of the first decades of the twentieth century, it is probable that many glaciers will disappear altogether. For example, in New Zealand Chinn (1988) calculates that if temperatures rise by 3.6–6.3°C, snow lines will migrate vertically by 300–500 m, and around 1000 out of the country's 3000 glaciers will disappear.

As we have seen, under most models of climatic change, the degree of temperature increase will be greatest in high latitudes. With a doubling of atmospheric CO_2 concentration a high latitude temperature rise of perhaps even as much as 10°C might be expected (Bell, 1982). Given that the existence and distribution of permafrost is so closely related to temperature, it is probable that major changes will occur. Barry (1985) estimates that an average northward displacement of the southern permafrost boundary by 150 ± 50 km would be expected for each 1°C warming, so that a total *maximum* displacement of between 1000 and 2000 km is possible.

However, major uncertainty surrounds the rate at which permafrost degradation will occur. As it is probably a slow process, permafrost will continue to exist for an extended time in areas of *continous* permafrost, in areas of discontinuous or sporadic permafrost the rate will vary greatly depending on local material conductivity, snow cover and vegetation. Changes in vegetation type and in snow cover in a warmer world may modify the direct consequences of warmer surface temperatures (Boer *et al.*, 1990).

Where the permafrost is ice rich, or contains massive ground ice segregations, subsidence and settling due to thawing will occur, inducing a thermokarst relief which can alter drainage patterns and stream courses. Coastal retreat will also gather momentum as permafrost degrades in coastal lowlands, and large areas may be inundated as depressions are lowered to below current sea levels because of thaw settlement. River banks and lake and reservoir shorelines might also become amenable to faster rates of erosion, while slopes could become less stable and the active zone become thicker.

Given that vegetation cover is a major control of the operation of geomorphological processes (Viles, 1988) it is inevitable that major changes in the nature and distribution of vegetation types will have a whole range of geomorphological consequences.

Some models suggest that wholesale changes in biomes will occur. Using the GCM developed by Manabe and Stouffer (1980) for a doubling of CO_2 levels, and mapping the present distribution of ecosystem types in relation to contemporary temperature conditions, Emmanuel *et al.* (1985) found *inter alia* that the following changes would take place: boreal forests would contract from their present position of comprising 23 per cent of total world forest cover to less than 15 per cent; grasslands would increase from 17.7 per cent of all world vegetation types to 28.9 per cent; deserts would increase from 20.6 to 23.8 per cent; and forests would decline from 58.4 to 47.4 per cent. More recently, Smith *et al.* (1992) have attempted to model the response of Holdridge's Life Zones to global warming and associated precipitation changes as predicted by a range of different GCMs. These are summarized for some major biomes in Table 11.1. All

Table 11.1 Changes in areal coverage (in km^2 × 10^3) of major biome as a result of the climatic conditions predicted for a × 2 CO_2 world by various General Circulation Models (GCMs)

Biome	*Current*	*OSU*	*GFDL*	*GISS*	*UKMO*
Tundra	939	−302	−5.5	−314	−573
Desert	3699	−619	−630	−962	−980
Grassland	1923	380	969	694	810
Dry forest	1816	4	608	487	1296
Mesic forest	5172	561	−402	120	−519

OSU = Oregon State University
GFDL = Geophysical Fluid Dynamic Laboratory
GISS = Goddard Institute for Space Studies
UKMO = UK Meteorological Office
Source: from Smith *et al.*, 1992.

the GCMs predict conditions that would lead to a very marked contraction in the tundra and desert biomes, and an increase in the areas of grassland and dry forests. There is, however, some disagreement as to what will happen to mesic forests as a whole, though within this broad class the humid tropical rainfall element will show an expansion.

Theoretically a rise of 1°C in mean annual temperature could cause a poleward shift of vegetation zones of about 200 km (Ozenda and Borel, 1990). However, uncertainties surround the question of how fast plant species would be able to move to and to settle new habitats suitable to the changed climatic conditions. Postglacial vegetation migration rates appear to have been in the range of a few tens of kilometres per *century.* For a warming of 2–3°C forest bioclimates could shift northwards about 4–6° of latitude in a century, indicating the need for a migration rate of some tens of kilometres per *decade.* Furthermore, migration could be hampered because of natural barriers, zones of cultivation, etc.

Altitudinal changes in vegetation zones will be of considerable significance. In general, Peters (1988) believes that with a 3°C temperature change vegetation belts will move about 500 m in altitude. One consequence of this would be the probable elimination of the Douglas Fir *(Pseudotsuga taxifolia)* from the lowlands of California and Oregon, because rising temperatures would preclude the seasonal chilling this species requires for seed germination and shoot growth.

Vegetation will also probably be changed by variation in the role of certain extreme events that cause habitat disturbance, including fire (exacerbated by drought), wind storms, hurricanes and coastal flooding (Overpeck *et al.*, 1990).

Other possible non-climatic effects of elevated CO_2 levels on vegetation include changes in photosynthesis, stomatal closure and carbon fertilization (Idso, 1983), though these are still matters of controversy (Solomon and West, 1986). However, especially at high altitudes, it is possible that elevated CO_2 levels would have potentially significant effects on tree growth, causing growth enhancement (La Marche *et al.*, 1984).

Changes in temperatures, precipitation quantities, and the timing and form of precipitation would have highly important hydrological consequences. In areas affected by snowfall today, the changes may be especially marked. In a warmer world, there would be a tendency for a marked decrease to occur in the proportion of winter precipitation that falls as snow. Furthermore, there would be an earlier and shorter spring snowmelt. The first of these two effects would cause greater winter rainfall and hence winter run-off, since less overall precipitation would enter snowpacks to be held over until spring snowmelt. The second effect would intensify spring run-off, leading to additional adverse consequences for both summer run-off levels and for spring and summer soil moisture levels (Gleick, 1986). Table 11.2, taken from Gleick's work, indicates likely changes in run-off for two temperature and five precipitation scenarios in such regions, and demonstrates clearly the anticipated direction and seasonality of run-off changes.

On the other hand, in high-latitude tundra environments warmer winters may cause more snow to fall, thereby creating increased run-off levels in the summer months (Barry, 1985). Indeed, Budyko (1982: 242) predicts that because of increased precipitation (perhaps by as much as 500–600 mm y^{-1} in the tundra zone), run-off in Russia north of 58–60°N will increase by a factor of 2 to 3.

Table 11.2 The effects of hypothetical temperature and precipitation scenarios on run-off (% change) (after Gleick, 1986)

	Precipitation change				
Temperature change	*−20%*	*−10%*	*0*	*+10%*	*+20%*
(a) *Winter (December, January, February)*					
+2°C	−24	−9	+8	+25	+44
+4°C	−4	+14	+34	+54	+75
(b) *Summer (June, July, August)*					
+2°C	−42	−32	−22	−12	−1
+4°C	−73	−68	−62	−55	−49

No less significant run-off changes may be anticipated for semi-arid environments, such as the South-West USA. For example, Revelle and Waggoner (1983) suggest that in the event of there being a 2°C rise in temperature and a 10 per cent reduction in precipitation, water supplies would be diminished by 76 per cent in the Rio Grande region and by 40 per cent in the Upper Colorado. Table 11.3 demonstrates that a 2°C rise in temperature would be

Table 11.3 Approximate percentage decrease in run-off for a 2°C increase in temperature

Initial temperature	*Precipitation (mm y⁻¹)*					
(°C)	200	300	400	500	600	700
−2	26	20	19	17	17	14
0	30	23	23	19	17	16
2	39	30	24	19	17	16
4	47	35	25	20	17	16
6	100	35	30	21	17	16
8		53	31	22	20	16
10		100	34	22	22	16
12			47	32	22	19
14			100	38	23	19

Source: data in Revelle and Waggoner, 1983.

most serious for water supplies and run-off in those regions where the mean annual precipitation is less than about 400 mm y^{-1}.

Projected summer dryness in areas such as the North American Great Plains (Manabe *et al.*, 1981) may be accentuated by a positive feedback process involving changes in cloud cover. Manabe and Wetherald (1986: 627) suggest that:

> When soil moisture is reduced, a large fraction of radiative energy absorbed by the continental surface is ventilated through the upward flux of sensible heat rather than through evaporation. Accordingly, the temperature of the continental surface and the overlying layer increases, resulting in the general reduction in relative humidity and precipitation in the lower troposphere of the model. Accompanying the reduction in relative humidity is a reduction in total cloud amount, causing an increase in solar energy reaching the continental surface. Thus the radiation energy absorbed by the continental surface also increases, raising the rate of potential evaporation.

Substantial hydrological changes may also occur in the catchments of the Great Lakes of North America because of changes in precipitation, snowmelt timing and evapotranspirational loss. Models produced by Croley (1990) and Hartmann (1990) suggest that for a doubling of CO_2 levels there may be a 23–51 per cent reduction in net basin supplies to all the Great Lakes, and that as a result lake levels will fall at rates ranging from 13 mm per decade (for Lake Superior) to 93 mm per decade (for Lake Ontario).

Projected sea-level rise

Among the most important geomorphological consequences of a global warming would be a worldwide rise in sea level. This would occur as a result of two distinct processes: the thermal expansion of the upper layers of the oceans; and the melting of land ice (US Department of Energy, 1985; Titus, 1986).

The anticipated rise of sea level over the next century is the subject of contention, largely because of uncertainties about the behaviour of Antarctic ice. In the 1980s there were expectations that sea level could rise by over 3.5 m by 2100. Now, however, there is a tendency to view such values as excessive and the Intergovernmental Panel on Climate Change (Houghton *et al.*, 1990:

279) concluded 'that a rise of more than 1 metre over the next century is unlikely.' Nonetheless, this is a rate 3 to 6 times faster than that experienced over the last 100 years (1 to 2 mm yr). The Panel also recognized that there were still large uncertainties associated with the future contributions of Antarctica and Greenland to sea-level rise.

The degree of sea-level rise could be moderated by reservoir construction. Newman and Fairbridge (1986) have calculated that between 1957 and 1982 human intervention stored as much as 0.75 mm per year of sea-level rise potential in reservoirs and irrigation projects.

The degree of tidal inundation may show local variations. Goemans (1986), for example, has pointed out that the tidal system of the North Sea is rather complex, with several amphidromic points (points of zero tidal range). These points might move as a result of sea level change and could therefore lead to a change in local tidal amplitudes and of the relative high-water level.

The effects of global sea-level rise will be compounded in those areas that suffer from local subsidence as a result of local tectonic movements, isostatic adjustments and fluid abstraction.

The results of sea-level rise in sensitive areas (*e.g.*, low-lying coasts, deltas, marshes) will be marked. Broadus *et al.* (1986), for example, calculate that were the sea level to rise by just 1 m in 100 years, 12–15 per cent of Egypt's arable land would be lost and 16 per cent of the population would have to be relocated. With a 3 m rise the figures would be a 20 per cent loss of arable land and a need to relocate 21 per cent of the population. In Bangladesh, a 1 m rise would inundate 11.5 per cent of the total land area of the state and affect 9 per cent of the population directly, while a 3 m rise in sea level would inundate 29 per cent of the land area and affect 21 per cent of the population. Many of the world's major conurbations might be flooded in whole or in part, and sewers and drains rendered inoperative (Kuo, 1986).

Salt marshes, including the mangrove swamps of the tropics, are potentially highly vulnerable in the face of sea-level rise, particularly in those circumstances where sea defences and other barriers prevent the landward migration of marshes as sea-level rises. However, salt marshes are highly dynamic features and in some situations may well be able to cope, even with quite rapid rises of sea level.

Reed (1990) suggests that salt marshes in riverine settings may receive sufficient inputs of sediment that they are able to accrete sufficiently rapidly to keep pace with projected rises of sea-level. Areas of high tidal range, such as the marshes of the Severn Estuary in England/Wales, are also areas of high sediment transport potential and may thus be less vulnerable to sea-level rise. Likewise, some vegetation associations, *e.g., Spartina* swards, may be relatively more effective than others at encouraging accretion, and organic matter accumulation may itself be significant in promoting vertical build-up of some marsh surfaces. For marshes that are dependent upon inorganic sediment accretion, increased storm activity and beach erosion which might be associated with the greenhouse effect could conceivably mobilize sufficient sediments in coastal areas to increase their sediment supply.

Marsh areas that may be highly prone to sea-level rise include areas of deltaic sedimentation where, because of sediment movement controls (*e.g.*, reservoir construction) or because of cyclic changes in the locations of centres of deposition, rates of sediment supply are low. Such areas may also be areas with high rates of subsidence. A classic example of this is portions of the Mississippi delta. Park *et al.* (1986) undertook a survey of coastal wetlands in the USA and suggested that sea level change could, by 2100, lead to a loss of between 22 and 56 per cent of the 1975 wetland area, according to the degree of sea-level rise that takes place.

One particular type of marsh that may be affected by anthropogenically accelerated sea-level rise is the mangrove swamp. As with other types of marsh the exact response will depend on local setting, sources and rates of sediment supply, and the rate of sea level rise itself. However, mangroves may respond rather differently to other marshes in that they are composed of relatively long-lived trees and shrubs which means that the speed of zonation change will be less (Woodroffe, 1990).

Another potentially susceptible environment is the coral reef (Stoddart, 1990). In the 1980s there were widespread fears that if rates of sea-level rise were high (perhaps 2 to 3 m or more by 2100) they would submerge whole atoll systems (*e.g.* the Maldives). However, with the reduced expectations for the degree of sea level rise that may occur, there has arisen a belief that coral reefs may survive and even prosper with moderate rates of sea level rise. Kinsey and Hopley (1991), for instance, believe that reef growth will be stimulated by the rising sea levels of a warmer world and that reef production could double in the next 100 years from around 900 to 1800 million tonnes per year. On the other hand, Buddemeier and Smith (1988) took a rather pessimistic view of the response of reefs to a rate of sea level rise of 15 mm per year over the next century, believing that sustained maxima of reef growth were generally less than 10 mm per year.

However, reef accretion is not the sole response of reefs to sea level rises, for reef tops are frequently surmounted by small islands (cays and motus) composed of clastic debris. Such islands might be very susceptible to sea level rise. On the other hand, were warmer seas to produce more storms, then the deposition of large amounts of very coarse debris could in some circumstances lead to their enhanced development.

Increased sea-surface temperatures could have deleterious consequences in corals which are near their thermal maximum, and increased temperatures in recent years have been identified as a cause of widespread coral bleaching (loss of symbiotic zooxanthellae). Those corals stressed by temperature or pollution might well find it more difficult to cope with rapidly rising sea levels than would healthy corals. Moreover, it is possible that increased ultraviolet radiation due to ozone layer depletion could aggravate bleaching and mortality caused by global warming. Various studies suggest that coral bleaching was a widespread feature in the warm years of the 1980s (Brown, 1990).

Bird (1986: 84) has summarized the general effects of accelerated coastal submergence as follows:

- On cliffed coasts submergence is likely to accelerate coastline recession, except on outcrops of hard rock formations, where the high and low tide lines will simply move up the cliff face. Existing shore platforms and abrasion ramps will disappear beneath the sea.

- The shores of deltas and coastal plains will retreat, except where they are maintained by coastal sedimentation.
- Beaches will be narrowed, and beach erosion will become much more extensive and severe than it is now.
- Inlets, embayments, and estuaries will be enlarged and deepened, and increasing salinity penetration will cause a regression of coastal ecosystems: where possible, mangrove and salt marsh communities will move back into terrain presently occupied by freshwater vegetation.
- Coastal lagoons will also become larger and deeper, but the enclosing barriers may transgress landward on to them. If the barriers are submerged, or destroyed by erosion, the lagoons will become coastal inlets or embayments.
- Low-lying areas on coastal plains, such as sebkhas (saline depressions now subject to occasional marine flooding) on arid coasts, will be flooded to form permanent lagoons.
- Upward growth of coral and associated organisms will be stimulated on fringing biogenic reefs, keeping pace with the marine transgression or lagging somewhat behind it.
- Erosion, structural damage, and marine flooding caused by storm surges or tsunamis will intensity because of the greater heights of the waves arriving through deepening coastal waters.
- Water tables will rise in coastal regions, and soil and water salinity will be augmented.

The relationship between sea-level rise and beach erosion is a notably important concern. In general, following the so-called Bruun Rule, such erosion is to be expected, but the precise amounts are difficult to quantify. However, on sandy beaches exposed to ocean waves, the coastline may erode by the order of 1 m for every 1 cm rise in sea level (Committee on Engineering Implications of Changes in Relative Mean Sea Level, 1987).

Environmental uncertainty

The debate about the relative power of human and natural influences is but one of many components of environmental uncertainty. In almost all attempts to predict the future major uncertainties arise. One can provide various explanations for this. First, environmental change is the result of very complex interactions between several closely coupled non-linear systems. The complexity creates problems for both modelling and comprehension, while the non-linearity means that the dimensions of a response are not by any means necessarily directly proportional to the size of the stimulus that promotes change. Secondly, prediction of environmental change depends on models and many models are imperfect because of the gross assumptions they employ (see the discussion of General Circulation Models for an exemplification of this problem). Thirdly, nature always has surprises in store (*e.g.,* various types of extreme event or catastrophe) and these will not cease in the future. They may work to counteract or increase the consequences of various human actions. Fourthly, there are bound to be factors that we have ignored, which turn out to be highly important. For example, who several decades ago could have predicted the role of CFCs in creating the ozone hole? Fifthly, identification of future trends requires some knowledge of background or natural levels. Often we lack the necessary long-term data to enable us to ascertain whether observed trends have happened before and whether they are or are not cyclical. Sixthly, some key issues are less easy to predict than others and therefore preclude accurate prediction of phenomena that depend on them. For example, without a clear idea of precipitation patterns in a warmer world it is well nigh impossible to predict the response of rivers and biota. Seventh, we may often be able to predict that change will probably occur, but we find it much less easy to predict the speed of response. Ice caps and glaciers may well melt in a warmer world, but how fast? Finally, there are problems of definition. Without clear definitions of phenomena, measurement is difficult, 'results' can be meaningless and trends or changes difficult to identify. We find it difficult to define such phenomena as desertification and deforestation.

□□□

12

Conclusion

The power of non-industrial and pre-industrial civilization

Thus far we have looked at the impact that human societies have had on the different components of the physical environment: vegetation, animals, soil, water, landforms and climate. It has become apparent that Marsh was correct over a century ago to express his cogently argued views of the importance of human agency in environmental change. Since his time the impact that humans have had on the environment has increased, as has our awareness of this impact. However, it is worth making the point here that, although much of the concern expressed about the undesirable effects humans have tends to focus on the role played by sophisticated industrial societies, this should not blind us to the fact that many highly significant environmental changes were and are being achieved by non-industrial societies. In recent years it has become apparent that fire, in particular, enabled early societies to alter vegetation substantially, so that plant assemblages that were once thought to be natural climatic climaxes may in reality be in part anthropogenic fire climaxes. This applies to many areas of both savanna and mid-latitude grassland. Such alteration of natural vegetation has been shown to pre-date the arrival of European settlers in the Americas, New Zealand and elsewhere and the effects of fire may have been compounded by the use of the stone axe and by the grazing effects of domestic animals. In turn the removal and modification of vegetation would have led to adjustments in fauna. It is also apparent that soil erosion resulting from vegetation removal has a long history and that it was regarded as a threat by the classical authors. Recent studies tend to suggest that some of the major environmental changes in highland Britain and similar parts of Western Europe that were once explained by climatic changes can be more effectively explained by the activities of Mesolithic and Neolithic peoples. This applies, for example, to the decline in the numbers of certain plants in the pollen record and to the development of peat bogs and podzolization. Even soil salinization started an early date because of the adaptation of irrigation practices in arid areas, and its effects on crop yields were noted in Iraq more than 4000 years ago. Similarly there is an increasing body of evidence that the hunting practices of early civilizations may have caused great changes in the world's mega-fauna as early as 11,000 years ago.

In spite of the increasing pace of world industrialization and urbanization, it is ploughing and pastoralism which are responsible for many of our most serious environmental problems and which are still causing some of our most widespread changes in the landscape. Thus soil erosion brought about by agriculture is, it can be argued, a more serious pollutant of the world's waters than is industry: many of the habitat changes which so affect wild animals are brought about through agricultural expansion; and soil salinization and desertification can be regarded as two of the most serious problems facing the human race. Land-use changes, such as the conversion of forests to fields, may be at least as effective in causing anthropogenic changes in climate as the more celebrated burning of fossil fuels and emission of industrial aerosols into the atmosphere. The liberation of CO_2 in the atmosphere through agricultural expansion, changes in surface albedo values and the production of dust, are all major ways in which agriculture may modify world climates.

The proliferation of impacts

A further point we can make is that, with developments in technology, the number of ways in which humans are affecting the environment is proliferating. It is these recent changes, because of the uncertainty which surrounds them and the limited amount of experience we have of their potential effects, which have caused greatest concern. Thus it is only since the Second World War, for example, that humans have had nuclear reactors for electricity generation, that they have used powerful pesticides such as DDT, and that they have sent supersonic aircraft into the stratosphere.

Likewise, it is only since around the turn of the century that the world's oil resources have been extensively exploited, that chemical fertilizers have become widely used, and that the internal combustion engine has revolutionized the scale and speed of transport and communications.

Above all, however, the complexity, frequency and magnitude of impacts is increasing, partly because of steeply rising population levels and partly because of a general increase in *per capita* consumption. Thus some traditional methods of land use such as shifting agriculture and nomadism, which have been thought to sustain some sort of environmental equilibrium, seem to break down and to cause environmental deterioration when population pressures exceed a particular threshold.

Fertility can be maintained under the long cycles characteristic of low-density populations. However, as population levels increase, the cycles necessarily become shorter, and soil fertility levels are not maintained, thereby imposing greater stresses on the land.

At the other end of the spectrum, increasing incomes, leisure and ease of communication, have generated a stronger demand for recreation and tourism in the developed nations. These have created additional environmental problems, especially in coastal and mountain areas (see, for example, Bayfield, 1973). Some of the environmental consequences of recreation can be listed as follows:

1. Desecration of cave formations by speleologists
2. Trampling by human feet leading to soil compaction
3. Nutrient additions at camp sites by people and their pets
4. Decrease in soil temperatures because of snow compaction by snowmobiles
5. Footpath erosion and off-road vehicle erosion
6. Dune reactivation by trampling
7. Vegetation change due to trampling and collecting
8. Creation of new habitats by cutting trails and clearing camp sites
9. Pollution of lakes and inland waterways by gasoline discharge from outboard motors and by human waste

10. Creation of game reserves and protection of ancient domestic breeds
11. Disturbance of wildlife by proximity of persons and by hunting, fishing and shooting
12. Conservation of woodland for pheasant shooting.

Likewise, it is apparent when considering the range of possible impacts of one major type of industrial development that they are significant. As Table 12.1 indicates, the exploitation of an oilfield and all the activities that it involves (for example, pipelines, new roads, refineries, drilling, etc.) have a wide range of likely effects on land, air, water and organisms.

A very substantial amount of change has been achieved in recent decades. Table 12.2, based on the work of Kates *et al.* (1991), attempts to make quantitative comparisons of the human impact on ten 'component indicators of the biosphere.' For each component they defined total net change clearly induced by humans to be 0 per cent for 10,000 years ago and 100 per cent for 1985. They then estimated dates by which each component had reached successive quartiles (*i.e.,* 25, 50 and 75 per cent) of its 1985 total change. They believe that about half of the components have changed more in the single generation since 1950 than in the whole of human history before that date.

Changes are reversible

It is evident that while humans have imposed many undesirable and often unexpected changes on the environment, they often have the capacity to modify the rate of such changes or to reverse them. There are cases where this is not possible: once soil has been eroded from an area it cannot be restored; once a plant or animal has become extinct it cannot be brought back; and once a laterite iron pan has become established it is difficult to destroy.

However, through the work of George Perkins Marsh and others, people became aware that many of the changes that had been set in train needed to be reversed or reduced in degree. Sometimes this has simply involved discontinuing a practice which has proved undesirable (such as the cavalier use of DDT), or replacing it with another which has been less detrimental in its

Table 12.1 Qualitative environmental impacts of mineral industries with particular reference to an oilfield

Facility	*Direction of the impact and reaction of the environmental*			
	Land	*Air*	*Water*	*Biocenosis*
Well	Alienation of land surface Extraction of oil associated gas, groundwater Pollution by crude oil, refined products, drilling mud Disturbance of internal structure of soil and subsoil Destruction of soil	Pollution by associated gas and volatile hydrocarbons, products of combustion	Withdrawal of surface water and groundwater Pollution by crude oil and refined products; salination of fresh water Disturbance of water balance of both subsurface and surface waters	Pollution by crude oil and refined products Disturbance and destruction over a limited surface area
Pipeline	Alienation of land Accidental oil spills Disturbance of landforms and internal structure of soil and subsoil	Pollution by volatile hydrocarbons	Disturbance and destruction over a limited surface area	
Motor roads	Alienation of land Pollution by oil products Disturbance of landforms and internal structure of soil and subsoil	Pollution by combustion products, volatile hydrocarbons, sulphur dioxide, nitrogen oxides	Pollution by combustion products Disturbances and destruction over limited surface area	
Collection point	Alienation of land Pollution by crude oil and refined products (spills) Disturbance of internal structure of soil and subsoil	Pollution by volatile hydrocarbons	Disturbance and destruction over limited surface area	

Source : Denisova, 1977, table 2. p. 650

Table 12.2 Chronologies of human-induced transformations

(a) Quartiles of change from 10,000 BC to mid-1980s

Form of transformation	*Dates of quartiles* 25%	50%	75%
Deforested area	1700	1850	1915
Terrestrial vertebrate diversity	1790	1880	1910
Water withdrawals	1925	1955	1975
Population size	1850	1950	1970
Carbon releases	1815	1920	1960
Sulphur releases	1940	1960	1970
Phosphorus releases	1955	1975	1980
Nitrogen releases	1970	1975	1980
Lead releases	1920	1950	1965
Carbon tetrachloride production	1950	1960	1970

(b) % change by the time of Marsh and Princeton symposium

Form of transformation	*% change* 1860	1950
Deforested area	50	90
Terrestrial vertebrate diversity	25–50	75–100
Water withdrawals	15	40
Population size	30	50
Carbon releases	30	65
Sulphur releases	5	40
Phosphorus releases	<1	20
Nitrogen releases	<1	5
Lead releases	5	50
Carbon tetrachloride production	0	25

Source: from Kates *et al*., 1991.

effect. Often, however, specific measures have been taken which have involved deliberate decisions of management and conservation.

Denson (1970) for example, outlines a sophisticated six-stage model for wildlife conservation:

1. Immediate physical protection from humans and from changes in the environment
2. Educational efforts to awaken the public to the need for protection and to gain acceptance of protective measures
3. Life-history studies of the species to determine their habitat requirements and the causes of their population decline
4. Dispersion of the stock to prevent loss of the species by disease or by a chance event such as a fire
5. Captive breeding of the species to assure higher survival of young, to aid research and to reduce the chances of catastrophic loss
6. Habitat restoration or rehabilitation when this is necessary before reintroducing the species.

Many conservation measures have been successful, while others have created as many problems as they were intended to solve. This applies, for example, to certain schemes for the reduction of coast erosion.

The concern with preservation and conservation has been longstanding, with many important landmarks. Interest has grown dramatically in recent years. Lowe (1983) has identified four stages in the history of British Nature Conservation:

1. The natural history/humanitarian period (1830–90)
2. The preservationist period (1870–1940)
3. The scientific period (1910–70)
4. The popular/political period (1960–present)

The first of these stages was rooted in a strong enthusiasm for natural history, and the crusade against cruelty to animals. Although many Victorian naturalists were avid collectors, numerous clubs were established to study nature and some of them sought to preserve species to make them available for observation. As we shall see, certain acts were introduced at this time to protect birds. During the preservationist period, there was the formation of a spate of societies devoted to preserving open land and its associated wildlife (for example, the National Trust, 1894; and the Council for the Preservation of Rural England, 1926). There was a growing sense of the vulnerability of wildlife and landscapes to urban and industrial expansion and geographers like Vaughan Cornish (see Goudie, 1972b) campaigned for the creation of national parks and the preservation of scenery, made possible through the National Parks and Access to Countryside Act of 1949. From the First World War onwards ecological research developed, and there arose an increasing understanding of ecological relationships. Scientists pressed for the regulation of habitats and species, and the Nature Conservancy was established in 1949. In the 1960s and the years that followed popular interest in conservation and widespread media attention first developed. This was partly generated by pollution incidents (such as the wrecks of the *Torrey Canyon* and *Amoco Cadiz),* and a gathering sense of impending environmental doom. Ecology became a political issue in various European nations, including the UK. In many countries major developments in land use, construction and industrialization now have to be preceded by the production of an Environmental Impact Assessment.

Thus in some countries, and in connection with particular species, conservation and protection have had a long and sometimes successful impact. In Britain, for example, the Wild Birds Protection Act dates back to 1880, and the Sea Birds Protection Act even further to 1869. The various acts have been modified and augmented over the years to outlaw egg-collecting, pole-trapping, plumage importation and the capture or possession of a range of species. The effectiveness of the different acts can be measured in real terms. Over the last 50 years no species have been lost as British breeding birds due to lack of protection—the only major loss has been the Kentish plover which was in any case on the edge of its range. Perhaps more importantly, several species have successfully recolonized Britain, the most celebrated being the avocet and the osprey. Today both are firmly established, together with other species lost in the nineteenth century: the black-tailed godwit, the goshawk and the bittern. Also as a result of protection the red kites of Wales have not only survived but increased in number, and the peregrine falcon maintains its largest numbers in Europe outside Spain.

One further ground which gives some basis for hope that humans may soon be reconciled with the environment is that there are some signs of a widespread shift in public attitudes to nature and the environment. These changing social values, combined with scientific facts, influence political action. This point of view, which acts as an antidote for some of the more pessimistic views of the world's future, has been elegantly presented by Ashby (1978). He contends that the rudiments of a healthy environmental ethic are developing, and explains:

> Its premise is that respect for nature is more moral than lack of respect for nature. Its logic is to put the Teesdale Sandwort . . . into the same category of value as a piece of Ming porcelain, the Yosemite Valley in the same category as Chartres Cathedral; a Suffolk landscape in the same category as a painting of the landscape by Constable. Its justification for preserving these and similar things is that they are unique, or irreplaceable, or simply part of the fabric of nature, just as Chartres and the painting by Constable are part of the fabric of civilization; also that we do not understand how they have acquired their durability and what all the consequences would be if we destroyed them.

Although there may be considerable controversy surrounding the precise criteria that can be used to select and manage sites that are particularly worthy of conservation (Goldsmith, 1983), there are nonetheless many motives behind the increasing desire to protect species and landscapes. These can be listed under the following general headings:

1. *The ethical.* It is asserted that wild species have a right to coexist with us on our planet, and that we have no right to exterminate them. Nature, it is maintained, is not simply there for humans to transform and modify as they please for their own utilitarian ends.
2. *The scientific.* We know very little about our surrounding environments, including, for example, the rich insect faunas of the tropical rain forest; therefore such environments should be preserved for future scientific study.

3. *The aesthetic.* Plants and animals, together with landscapes, may be beautiful and so enrich the life of humans.
4. *The need to maintain genetic diversity.* By protecting species we maintain the species diversity upon which future plant and animal-breeding work will depend. Once genes have been lost, they cannot be replaced.
5. *Environmental stability.* It is argued that in general the more diverse an ecosystem is the more checks and balances there are to maintain stability. Thus environments that have been greatly simplified by humans may be inherently unstable, and prone to disease, etc.
6. *Recreational.* Preserved habitats and landscapes have enormous recreational value, and in the case of some game reserves and natural parks may have economic value as well (for example, the safari industry of East Africa).
7. *Economic.* Many of the species in the world are still little known, and there is the possibility that we have great storehouses of plants and animals which, when knowledge improves, may become useful economic resources.
8. *Future generations.* One of the prime arguments for conservation, whether of beautiful countryside, rare species, soil or mineral resources, is that future generations (and possibly ourselves later in life) will require them, and may think badly of a generation that has squandered them.
9. *Unintended impacts.* As we have seen so often in this book, profligate or unwise actions can lead to side effects and consequences that may be disadvantageous to humans.

Some of these arguments are more utilitarian than others (for example, 4, 5, 6, 7 and 9), and some may be subject to doubt —it could, for example, be argued that future generations will have technology to use new resources and may not need some of those we regard as essential—but overall they provide a broadly based platform for the conservation ideal (Myers, 1979).

The susceptibility to change

Ecosystems respond in different ways to the human impact, and some are more vulnerable to human perturbation than others.

It has often been thought, for example, that complex ecosystems are more stable than simple ones. Thus in Clements' Theory of Succession the tendency toward community stabilization was ascribed in part to an increasing level of integration of community functions. As Goodman (1975: 238) has expressed it:

> In general the predisposition to expect greater stability of complex systems was probably a combined legacy of eighteenth century theories of political economics, aesthetically and perhaps religiously motivated attraction to the belief that the wondrous variety of nature must have some purpose in an orderly world, and ageless folkwisdom regarding eggs and baskets.

Indeed, as Murdoch (1975) has pointed out, it makes good intuitive sense that a system with many links, or 'multiple fail-safes', is more stable than one with few links or feedback loops. As an example, if a type of herbivore is attacked by several predatory species, the loss of anyone of these species will be less likely to allow the herbivore to erupt or explode in numbers than if only one predator species were present and that single predator type disappeared.

Various other arguments have been marshalled to support the idea that great diversity and complexity affords greater ability to minimize the magnitude, duration and irreversibility of changes brought about by some external perturbation such as humans (Noy-Meir, 1974). It has been stated that natural systems, which are generally more diverse than artificial systems such as crops or laboratory populations, are also more stable. Likewise, the tropical rain forest has been thought of as more diverse and more stable than less complex temperate communities, while simple Arctic ecosystems, exemplified by unstable lemming populations, are in turn held to be less stable than the more diverse temperate communities. In the same way the simple ecosystems of oceanic islands have always appeared highly vulnerable to disturbance brought about by anthropogenic plant and animal introductions.

However, considerable doubt has been expressed recently as to whether the classic concept of the causal linkage between diversity/ complexity and stability is entirely valid (see, for example, Hurd *et al.*, 1971). Murdoch (1975) indicates that there is no convincing field evidence that diverse natural communities are generally more stable than simple ones. He cites various papers which show that fluctuations of microtine rodents (lemmings, field voles, etc.) are as violent in relatively complex temperate zone ecosystems as they are in the less complex Arctic zone ecosystems. This is supported by Goodman (1975: 239) who writes:

> As for the apparent stability of tropical biota, that could well be an illusion attributable to insufficient study of bewilderingly complex assemblages in which many species are so poorly represented in samples of feasible size that even considerable fluctuations might go undetected. Indeed, there are countervailing anecdotes regarding ecological instability in the tropics, such as recent reports of an insect virtually defoliating the wild Brazil-nut trees in Bolivia and of monkeys succumbing in large numbers to epidemics.

He goes on to add: 'There is growing awareness of the surprising susceptibility of the rain forest ecosystem to man-made perturbation.' This is a point of view supported by May (1979) and discussed by Hill (1975). Hill points out that a very high species diversity is frequently associated with areas which have relatively constant physical environmental conditions over the course of a year and a series of years. The rain forest may be construed to be such an environment, and one where this constancy has allowed the presence of many specialized species, each pursuing a narrow range of activities. It has been argued that because of the high degree of specialization, the indigenous species have a limited ability to recover from major stresses caused by human intervention.

Goodman (1975) has also queried the sufficiency of the argument in its reference to the apparent instability of island ecosystems, suggesting that islands, being evolutionary backwaters and dead-ends, may accumulate species that are especially susceptible to competitive or exploitative displacement. In this case, lack of diversity may not necessarily be the sole or prime cause of instability.

The apparent instability of agricultural compared to natural communities is also often attributed to lack of diversity and indeed modern agriculture does involve significant ecosystem simplification. However, such instability as there is may not, once again, necessarily result from simplification. Other factors could promote instability: agricultural communities are disrupted, even destroyed, more frequently and more massively as part of the cultivation process than those natural systems we tend to think of as stable; the component species of natural systems are co-evolved (co-adapted), and this is not usually true of agricultural communities. As Murdoch (1975: 799) suggests, it may be that:

> Natural systems are more stable than crop systems because their interacting species have had a long shared evolutionary history.... In contrast with these natural communities the dominant plant species of a crop system is thrust into an often alien landscape... the crops have undergone radical selection in breeding programmes, often losing their genetic defence mechanisms.

Thus the ideas that complex natural ecosystems will be less susceptible to human interference and that simple artificial ecosystems will inevitably be unstable are not necessarily tenable. Nonetheless, it is apparent that there are differences in susceptibility between different ecosystem types, and these differences may result from factors other than the degree of diversity and complexity (Cairns and Dickson, 1977).

Some systems tend to be *vulnerable*. Lakes, for example, are natural traps and sumps and are thus more vulnerable to the effects of disadvantageous inputs than are rivers (which are continually receiving new inputs) or oceans (which are so much larger). Other systems display the property of *elasticity*—the ability to recover from damage. This may be because nearby epicentres exist to provide organisms to reinvade a damaged systems. Small, isolated systems will often tend to possess low elasticity. Two of the most important properties, however, are *resilience* (being a measure of the number of times a system. can recover after stress), and *inertia* (the ability to resist displacement of structural and functional characteristics).

Two systems which display resilience and inertia are deserts and estuaries. In both cases their indigenous organisms are highly accustomed to variable environmental conditions. Thus most desert fauna and flora evolved in an environment where the normal pattern is one of more or less random alternations of short favourable periods and long stress periods. They have pre-adapted resilience (Noy-Meir, 1974) so that they can tolerate extreme conditions, have the ability for rapid recovery, have various delay and trigger mechanisms (in the case of plants) and have flexible and opportunistic eating habits (in the case of beasts). Estuaries, on the other hand, though the subject of increasing human pressures, also display some resilience. The vigour of their water circulation continuously and endogenously renews the supply of water, food, larvae, etc.; this aids recovery. Also, many species have biological characteristics which provide special advantages in estuarine survival (Cronin, 1967). These characteristics usually protect the species against the natural violence of estuaries and are often helpful in resisting external forces such as humans.

Human influence or nature?

From many of the examples given in this book it is apparent that in many cases of environmental change it is impossible to state, without risk of contradiction, that people rather than nature are responsible. Most systems are complex and human agency is but one component of them, so that many human actions can lead to end-products which are intrinsically similar to those that may be produced by natural forces. It is a case of equifinality, whereby different processes can lead to basically similar results. Humans are not always responsible for some of the changes with which they are credited. This book has given many examples of this problem and a selection is presented in table 12.3. Deciphering the cause is often a ticklish problem, given the intricate interdependence of different components of ecosystems, the frequency and complexity of environmental changes, and the varying relaxation times that different ecosystem components may have when subject to a new impulse. This problem plainly docs not apply to the same extent to changes which have been brought about deliberately and knowingly by humans, but it does apply to the many cases where humans may have initiated change inadvertently and unintentionally.

Table 12.3 Human influence or nature? Some examples with page references to this book

	Causes		
Change	*Natural*	*Anthropogenic*	*Page reference*
Late Pleistocene animal extinction	Climate	Hunting	125
Death of savanna trees	Soil salinization through climatically induced groundwater rise	Overgrazing	–
Desertification in semi-arid areas	Climatic change	Overgrazing, etc.	55
Holocene peat-bog development in highland Britain	Climatic change and progressive soil deterioration	Deforestation and ploughing	151
Holocene elm and linden decline	Climatic change	Feeding and stalling of animals	65
Tree encroachment into alpine pastures in USA	Temperature amelioration	Cessation of burning	–
Gully development	Climatic change	Land-use change	263
Early twentieth-century climatic warming	Changes in solar emission and volcanic activity	CO_2-generated greenhouse effect	304
Increasing coast recession	Rising sea-level	Disruption of sediment supply	286
Increasing coastal flood risk	Rising sea-level, natural subsidence	Pumping of aquifers creating subsidence	259
Increasing river flood intensity	Higher intensity rainfall	Creation of drainage ditches	254
Ground collapse	Karstic process	Dewatering by over-pumping	79
Forest decline	Drought	Air, soil and water pollution	–

This fundamental difficulty means that environmental impact statements of any kind are extremely difficult to make. As we have seen, humans have been living on the earth and modifying it in different degrees for several millions of years, so that it is problematical to reconstruct any picture of the environment before human intervention. We seldom have any clear baseline against which to measure changes brought about by human society. Moreover, even without human interference, the environment would be in a perpetual state of flux on a great many different timescales (Goudie, 1992). In addition, as Wall and Wright (1977: 3) have demonstrated, there are spatial and temporal discontinuities between cause and effect. For example, erosion in one locality may lead to deposition in another, while destruction of key elements of an animal's habitat may lead to population declines throughout its range. Likewise, in a rime context, a considerable interval may elapse before the full implications of an activity are apparent. Also, because of the complex interactions between different components of different environmental systems and subsystems it is almost impossible to measure total environmental impact. For example, changes in soil may lead to changes in vegetation, which in turn may trigger changes in water quality and in animal populations. Primary impacts give rise to a myriad of successive repercussions throughout ecosystems which may be impracticable to trace and monitor. Quantitative cause-and-effect relationships can seldom be established.

Into the unknown

During the 1980s and 1990s the full significance of possible future environmental changes has become apparent, and national governments and international institutions have begun to ponder whether the world is entering a spasm of unparalleled humanly induced modification. Our models and predictions are still highly inadequate, and there are great ranges in some of the values we give for such crucial changes as sea level rise and global climatic warming, but the balance of scientific argument favours the view that change will occur and that change will be substantial. Some of the changes may be advantageous for humans or for particular ecosystems; others will be extremely disadvantageous. But all change, if it is rapid and of a great magnitude, is likely to create uncertainties and instabilities. The study of future events will not only become a focus for the environmental sciences, but will also become a major concern for economists, sociologists, lawyers and political scientists. George Perkins Marsh was a lawyer and politician, but it is only now, a century and a quarter since he wrote *Man and Nature,* that the wisdom, perspicacity and prescience of his ideas have begun to be given the praise and attention they deserve.

INDEX